A Textbook
of
Solid State Chemistry

A Textbook
of
Solid State Chemistry

Dr. Mahima Srivastava
[M.Sc. (Gold Medalist), Ph.D.]
Associate Professor,
D.B.S. (PG) College, Dehradun,
H.N.B. Garhwal Central University,
Srinagar Garhwal, Uttarakhand, India

CWP
Central West Publishing

I dedicate this book to my parents,
brother Anuj Anand and son Daksh Srivastava

Foreword

Our world in present times is conceived more of being molecular or materials oriented. Solid-state chemistry is an area of increasing importance. Most of the technology finds its basics in solid-state chemistry. The switches, sensors, rectifiers, transistors, superconductors, optical storage devices are the talk of the current times.

Thus, the importance of solid-state chemistry continues to increase and the book by Dr. Mahima Srivastava is a welcome addition to the available literature on the subject. Its scope places it in a useful niche among all the learners of this category. The author has retained the tested and tried classical approach but has successfully placed her own individual spin on her arrangement. She has put together a selected range amongst the most important of the vast array of content available. The matter is ordered in a clear, logical, and lucid pattern over the whole book. The book includes content dealing with basic aspects of metals such as defect structures in metal oxides of non-stoichiometric composition their transport kinetics, formation of spinels and organic metals. Applied aspects concern the organic metals, fullerenes and doped fullerenes.

The present work should be of value to the students and practitioners of solid-state chemistry at all levels. It is also recommended to researchers working in the fields of solid-state chemistry and students in the field of materials science. It will be of particular assistance in presenting a crisp and modern view of the subject to those who use solid-state chemistry.

I heartily congratulate Dr. Mahima Srivastava for the magnificent job she has done.

C.N.R. Rao
Bharat Ratna
Jawaharlal Nehru Centre for Advanced Scientific Research

Contents

Preface

I feel immense pleasure in presenting this book to the teachers and students of undergraduate and postgraduate level. This book is the outcome of my long teaching experience of more than a decade at graduate and postgraduate level. A long span of interaction with students has guided me to give special attention to dark areas which students find difficult to understand. The book has been written in a simple and easy-to-understand language. Some salient features of the book are:

- Many illustrations and examples have been provided to make the reading interesting and engaging.
- A large number of synthesis methods and reactions for every compound have been discussed.
- The compounds in every category have been discussed in detail for clear understanding.
- General methods of synthesis and reactions has also been elaborated along with specific methods of synthesis and reactions so that basic pattern in which the compound behaves may be followed.
- Various images have been added for pictorial understanding of the topic.

I have freely consulted edited scientific work such as Solid-State Chemistry and Its Applications edited by A. R. West and a wide variety of research papers available on the related subject. I am very grateful to the editors, authors and publishers of these works.

I am thankful to the management and editorial team of Central West Publishing for the help and support in the publication of this book.

I believe that the book contains all that is needed to understand the subject in a systematic manner. However, I also believe that no work is perfect, and there is always scope for improvement. I would welcome suggestions from the teaching fraternity and students for further improvement of the book.

Mahima Srivastava

Symbols and Abbreviations

LFSE	ligand field stabilization energy
OSSE	octahedral site stabilization energy
CFSE	crystal field stabilization energy
$Ohm^{-1}cm^{-1}$	conductivity
B	magnetic induction
H	magnetic field
I	moment per unit volume
μ	magnetic permeability
T_c	Curie temperature
Fcc	face centered cubic
Bcc	body centered cubic
MRI	magnetic resonance imagery
ED	electron donor
EA	electron acceptor
TCNQ	tetracyanoquinodimethane
TTF	tetrathiafulvalene
DAP	2,3-diaminopyridine
CLA	chloranilic acid
HTS	high temperature superconductors
LHC	Large Hadron Collider
SQUID	superconducting quantum interface devices
IFLM	inorganic fullerene like material
TDAE	tetrakis (dimethylamino) ethylene
HOMO	highest occupied molecular orbital
LUMO	lowest unoccupied molecular orbital
THCT	tunneling hot carrier transistor
MIMIM	metal-insulator-metal-insulator-metal
WORM	write once read many
WMRM	write many read many
SSR	solid-state relay
TRIAC	TRIode for alternating current
MOSFET	metal oxide semiconductor field effect transistor
LDR	light dependent resistor

Chapter 1

Introduction

From our past knowledge we know that matter exists in three states – **SOLID, LIQUID and GAS.** Liquids and gases have a property of flowing as their molecules are free to move and hence are called **FLUIDS**. But solids, due to intactness in their composition and oscillation across their mean position, deny flexibility in movement and are rigid in property. Due to the same reason, they have definite shape, volume and mass. There are many more properties of solids which find use due to modification like the solid surface opposes any force that is applied on the surface. They are also incompressible due to the closed arrangement of their constituent particles.

Material chemistry, also known as solid state chemistry, is the science of a variety of solid materials and their structure, synthesis, properties and applications. The science also facilitates the understanding of molecular level to crystal structure level of the compound. Non-stoichiometric compounds are also placed in solid state compounds due to their thermodynamic properties and structure. On the basis of atomic arrangement, binding energy, physical & chemical properties solids are classified as: **AMORPHOUS** and **CRYSTALLINE**.

Amorphous solid states are those states of solids which have restricted rigidity and incompressibility. Examples of amorphous solids are plastics glasses, pitch, etc. Amorphous solids can also be liquids that have been super cooled, and it is due to this reason that they tend to flow although very slowly (in negligible manner). They have a capability of being molded into desired shapes as they can moderately soften over a range of temperature. Amorphous solids lack regular shape which means that there is undefined geometry in the arrangement of respective particles. These solids show **isotropy** i.e., due to irregular arrangement the value of any physical quantity would remain same along any direction.

The state of solid having defined geometry is the crystalline solid state. These solids possess a regular pattern of arrangement of the constituent particles throughout the three-dimensional network. Examples of crystalline solids are diamond, common salt, quartz, etc.

1

There is found geometrical shape in such compounds with flat faces and sharp edges which is termed as crystalline lattice. They have definite and sharp melting points. Such compounds tend to be **anisotropic** meaning there are different values for various physical properties being measured in different directions for the same crystal. Thus, crystalline solids are actual solids.

There have been identified **four** types of crystalline solid states viz., **ionic solids** (comprising of ions), **metallic solids** (comprising of positive ions in the cluster of free electrons), **molecular solids** (comprising of molecules) and **covalent solids** (comprising of covalently bonded atoms). Ionic solids have extraordinarily strong electrostatic force between the ions hence they are hard, brittle and insulators in solid state; however, they are conductors in molten or solution state. Metallic solids have free electrons to allow electrical and thermal conductivity. These free electrons also tend to impart lustre, malleability and ductility to metallic solids.

Molecular Solids may be divided into three types: non–polar molecular solids, in which atoms/molecules are held by London forces; polar molecular solids, in which atoms/molecules are held by dipole–dipole interactive forces and hydrogen bonded molecular solids, which have covalent bonds that are polar in nature. Covalent or network solids are huge molecules which have directional bonds. They have higher order melting points and are brittle in nature.

Apart from amorphous and crystalline states, there are **polycrystalline solids** which are amorphous by appearance but microcrystalline by structure. Many crystallites of different size and orientation form the constituent particles of polycrystalline Solids, for example inorganic solids.

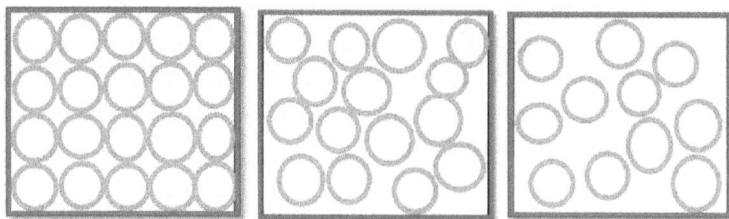

Figure 1.1 Crystalline (left), polycrystalline (middle) and amorphous (right) states.

The microscopically small **crystallites or grains** are formed during cooling of many materials. Crystallites do not have any recommended direction i.e., they are random. The grains are separated by grain boundaries.

There is an important area in the solids and that is of defects. The defects are a part of all solids and have an especially important impact on the properties of the solid.

Chapter 2

Definite Principles of Solid-State Reactions

Introduction

A mixture of various solids, to synthesize powders is the most used method for their preparations. They need to be heated at remarkably high temperatures to react as at normal room temperature they are appreciably unreactive. This means that for their active mixing and producing polycrystalline solids, factors involving change in free energy (thermodynamic) and factors involving rate of reaction (kinetic), both are important. We shall see this with the help of **spinel**, which are compounds with general formula XY_2Z_4 in which X and Y are cationic in nature coordinated tetrahedrally and octahedrally and Z is anionic in nature.

The common structure got its name from mineral **magnesium aluminate** having general composition, AB_2O_4 which is made up of oxides AO and B_2O_3 in equimolar ratios. In the spinel AB_2O_4, A is a divalent metal ion, B is a trivalent metal ion and Oxide ions are arranged in cubic closed packed lattice. At present, there are many compounds in this category as, galaxite [$MnAl_2O_4$], hercynite [$FeAl_2O_4$], gahnite [$ZnAl_2O_4$] are all aluminium spinels; jacobsite [$MnFe_2O_4$], trevorite [$NiFe_2O_4$], cuprospinel [$CuFe_2O_4$] are iron spinels; chromite [$FeCr_2O_4$] and zincochromite [$ZnCr_2O_4$] are chromium spinels; coulsonite [FeV_2O_4] is a vanadium spinel.

Under ordinary conditions of experiment, there occurs no change/product formation when these two oxides are mixed; however, at higher temperature they tend to respond as example, the formation of spinel $MgAl_2O_4$ could be achieved at temperatures around 1000 – 1200 °C and the amount of spinel phase would steadily increase with higher temperatures. The structure of spinel does not have complete identical structure to any of the constituents as shown below.

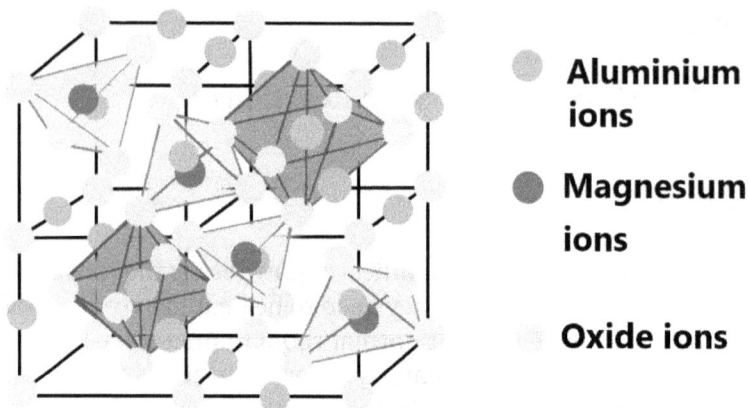

Figure 2.1 Cubic closed packed structure of MgAl$_2$O$_3$.

As can very well be seen from the structures above, that **magnesium aluminate spinel** has a cubic closed packed arrangement of oxide ions but in Al$_2$O$_3$, the same arrangement is of a deformed closed packed. These formations are not easy to bring about due to lack of a medium because of which formation and movement of ions becomes a difficult task to accomplish. Also, there is a lack of homogeneity due to difference in structures of the solids that are mixed. Since different ions are trapped in different lattice structures of the reactants, hence for the formation of the product, the old bonds need disintegration which becomes difficult at ordinary temperature.

The formation of **zinc ferrite spinel** occurs by mixing zinc oxide and ferric oxide and at appropriate temperature, first the nucleus of spinel is formed at the interface. Then, slowly in due course, the high temperature helps in breaking and making of bonds so that the spinel grows and more of product is formed with counter diffusion of ions.

As the product layer increases, the otherwise slow reaction turns to be even slower.

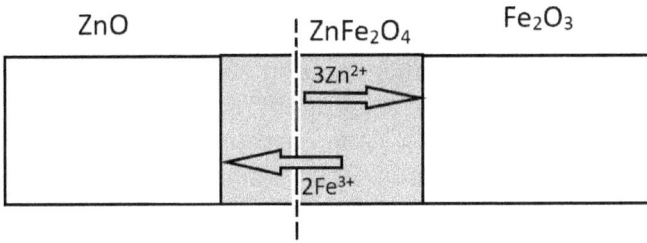

Figure 2.2 Formation of ZnFe$_2$O$_4$ spinel.

2.1 Wagner Reaction Mechanism

The formation of a spinel proceeds via **Wagner reaction mechanism**. The mechanism has been named after Carl Wilhelm Wagner who was a German chemist and is well known for his work on solid state reactions, counter diffusion of ions, and defect chemistry. He is also known as the father of solid-state chemistry. The Wagner mechanism proceeds by diffusion of counterions through the product layer. If A is divalent and B is trivalent, then three A^{++} ions will diffuse to maintain the charge against two B^{+++}.

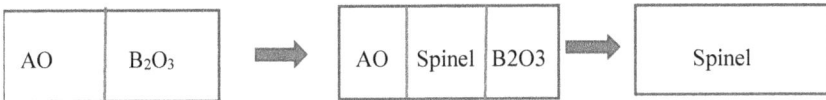

Figure 2.3 Procedural formation of a spinel.

This can well be understood from the reaction mechanism for the formation of ZnFe$_2$O$_4$ as:

1. **Reaction at ZnO/ZnFe$_2$O$_4$**
   $2Fe^{+++}$ - $3Zn^{++}$ + $4ZnO$ = $ZnFe_2O_4$

2. **Reaction at ZnFe$_2$O$_4$/Fe$_2$O$_3$**
   $3Zn^{++}$ - $2Fe^{+++}$ + $4Fe_2O_3$ = $3ZnFe_2O_4$

3. **Summary of 1 and 2**
   $4ZnO$ + $4Fe_2O_3$ = $4ZnFe_2O_4$

2.2 Kirkendall Effect

It is evident from above, that reaction (**2**) gives three times more product than reaction (**1**). If the color of reactants and product is different, then the formation of spinel becomes clearly visible and this effect, where the compounds participating in the reactions can be used as indicators of progress of reaction is referred to as **Kirkendall effect**. Thus, for a successful solid-state reaction,

1. the reactants should be finely powdered.
2. the surface area of contact of reactants should be maximum.
3. the temperature should be high enough to initiate nucleation of spinel.

Nucleation is the start of formation of the product i.e., the spinel and once the spinel has started to form then it is followed by its subsequent growth in size. The start of the product formation is favored if there is similarity in structure if reactants and product, for example, for the formation of magnesium aluminate spinel, nucleation is supported since there is similarity in the positioning of oxide ions in both spinel as well as magnesium oxide. Such reactions where structural similarity at the reaction site as well as through the entire reaction mixture is required for the ease of reaction to occur are called **topotactic reactions**.

As has been discussed earlier, the formation of spinel takes place due to diffusion of ions of the reactants; this process is greatly enhanced if defects are present in the crystal. The crystals, whether long order arrangement or short order arrangement, are not perfect. A combination of many small sized crystals makes a solid. Since these crystals are formed by the slow process of crystallization, hence they are liable to incur some defects. There are two types of defects: **point defects** and **line defects**.

The irregularity of crystalline structure around an atom (point) is termed as point defect. They are placed in three classes as:

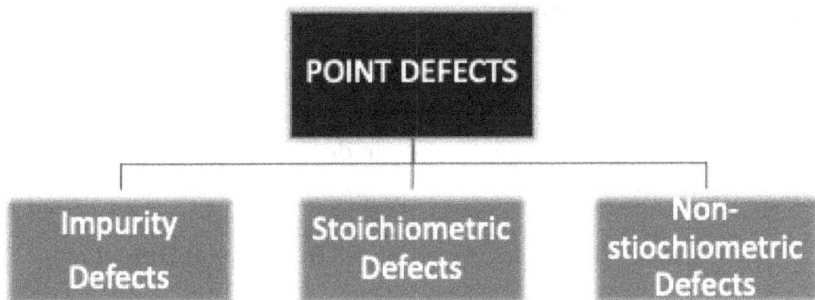

Figure 2.4 Classification of point defects.

2.3 Impurity Defects

Impurity defects can be understood with the example of a salt AB, where A is the positive ion and B is the negative ion. If AB, in molten state is mixed with small quantity of ZB_2, where Z is a divalent positive ion, and crystallized, then the impurity of Z^{++} occupies some sites of A^+ while some sites remain vacant. For example, in a mixture of NaCl and $SrCl_2$, one Sr^{++} ion sits at Na^+ seat while one Na^+ site remains vacant, hence overall charge remains unchanged but there occurs a defect in structure.

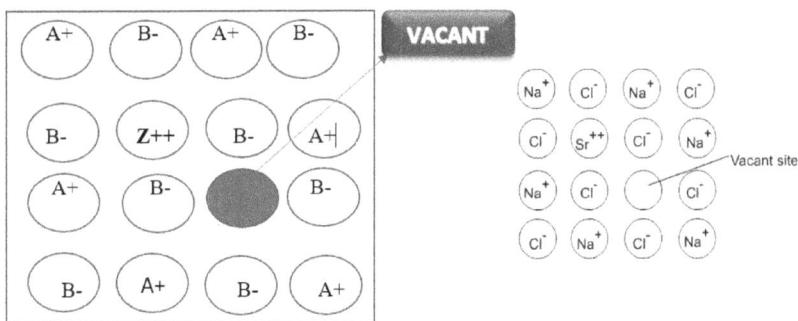

Figure 2.5 Impurity defect in A^+B^- and NaCl.

9

2.4 Stochiometric Defects

Stoichiometric defects are also called Intrinsic or thermodynamic defects and in these defects stoichiometry of the solid remains intact. There are two types of stoichiometric defects viz., **vacancy defects** and **interstitial defects**.

Vacancy Defect	Interstitial Defect
Contains vacant lattice sites in the crystal	Some atoms/molecules occupy extra space in the crystal
Results in decrease in the density of the compound	Results in increase in the density of the compound

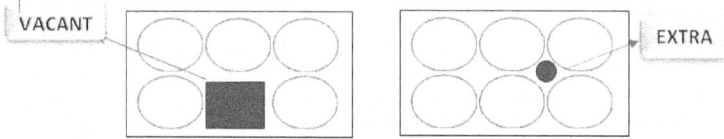

Since vacancy and interstitial defects are found in combination hence, according to their practical occurrence they are **Frenkel defects and Schottky defects**. The cation in ionic solids leaves its original site and settles at an interstitial position giving rise to Frenkel Defect. Schottky defect decreases the density of the substance since there are vacant sites in the ionic solids. Schottky defect is usually shown by those ionic solids which possess same sized cations and anions.

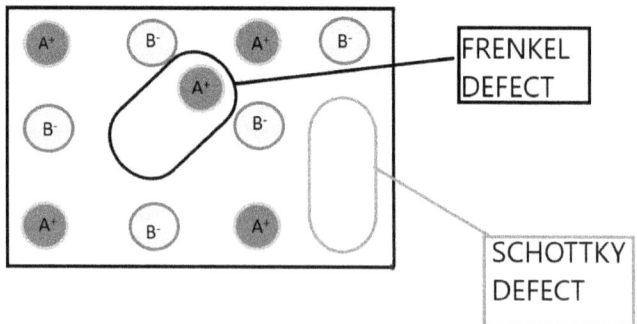

Figure 2.6 Frenkel and Schottky defects.

10

2.5 Non-stoichiometric Defects

There are certain defects in solids which bring about a change in the stoichiometry and hence are called **non-stoichiometric defects**.

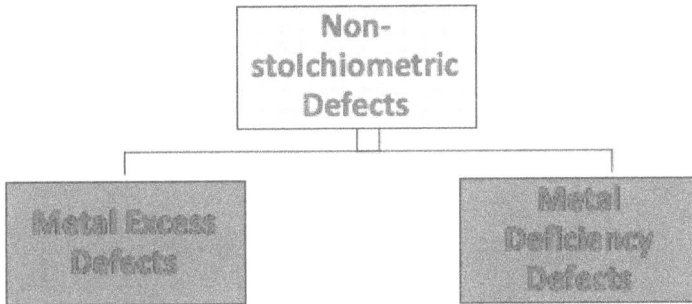

Metal Excess Defects

The defects which arise due to the presence of extra cations at interstitial sites or due to the absence of anions are called a metal excess defect. This type of defect is generally shown by alkyl halides like NaCl, KCl, etc.

These are of two types:

Due to anionic vacancies: A negative ion may be missing from its lattice site, leaving a hole that is occupied by an electron to maintain electrical neutrality.

Example: When crystals of NaCl are heated in an atmosphere of sodium vapours, the sodium atoms are deposited on the surface of the crystal. The Cl- ions diffuse to the surface and combine with Na+ ions to give NaCl. Na+ is formed by loss of an electron by Na atom. The released electron diffuses into the crystal and occupy anionic sites. This anionic site occupied by the unpaired electron is called F – center (FARBENZENTER). The F – center imparts a yellow color to the crystals of NaCl. The color occurs due to excitation of these electrons when they absorb energy from the visible light falling on the crystals. Likewise, excess lithium makes LiCl crystals pink and excess of potassium makes KCl crystals lilac.

Due to the presence of extra cation: Extra cations are present at the interstitial sites and the electrons at another interstitial site maintain electrical neutrality.

Example: Zinc oxide is white in color at room temperature, on excess heating, it turns yellow as it loses oxygen.

$ZnO = Zn^{2+} + \frac{1}{2}O_2 + 2e^-$ **[on heating]**

Now there is excess of zinc ions in the crystal which move to the interstitial sites and electrons to the neighboring interstitial sites.

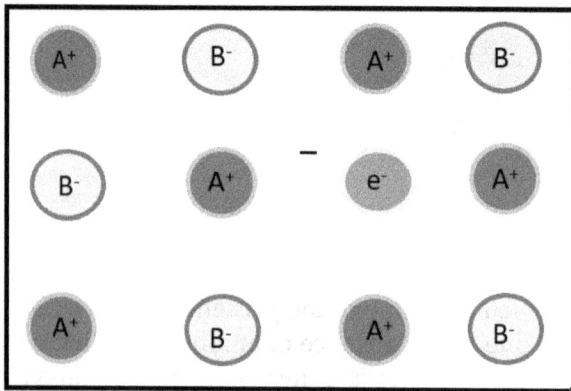

(a)

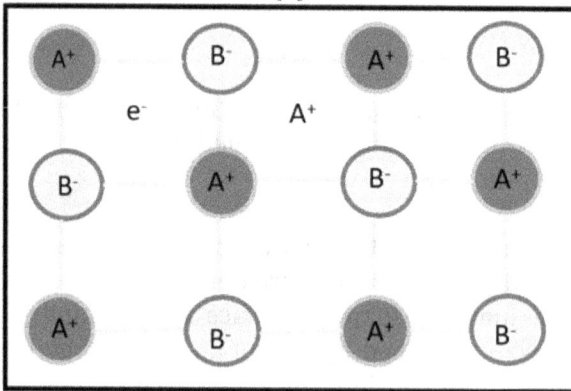

(b)

Figure 2.7 Metal defects (a) due to anionic vacancy (b) due to interstitial anion.

Metal Deficient Defects

The defect which arises when the metal shows variable valency.
Types of metal deficient effect- These are of two types:

Due to cation vacancies: A positive ion may be missing from its lattice site, and the extra negative charge balances by acquiring two positive charges instead of one. This type of defect usually occurs in compounds having a variable oxidation state.

Example: Ferrous oxide, Nickel oxide. In the crystals of FeO, some Fe^{2+} cations are missing, and the loss of positive charge is made up by the presence of required number of Fe^{3+} ions.

Due to the presence of extra anion: Extra anions are present at the interstitial sites and the adjacent ions at another interstitial site maintain electrical neutrality. This type of defect is found in rare cases.

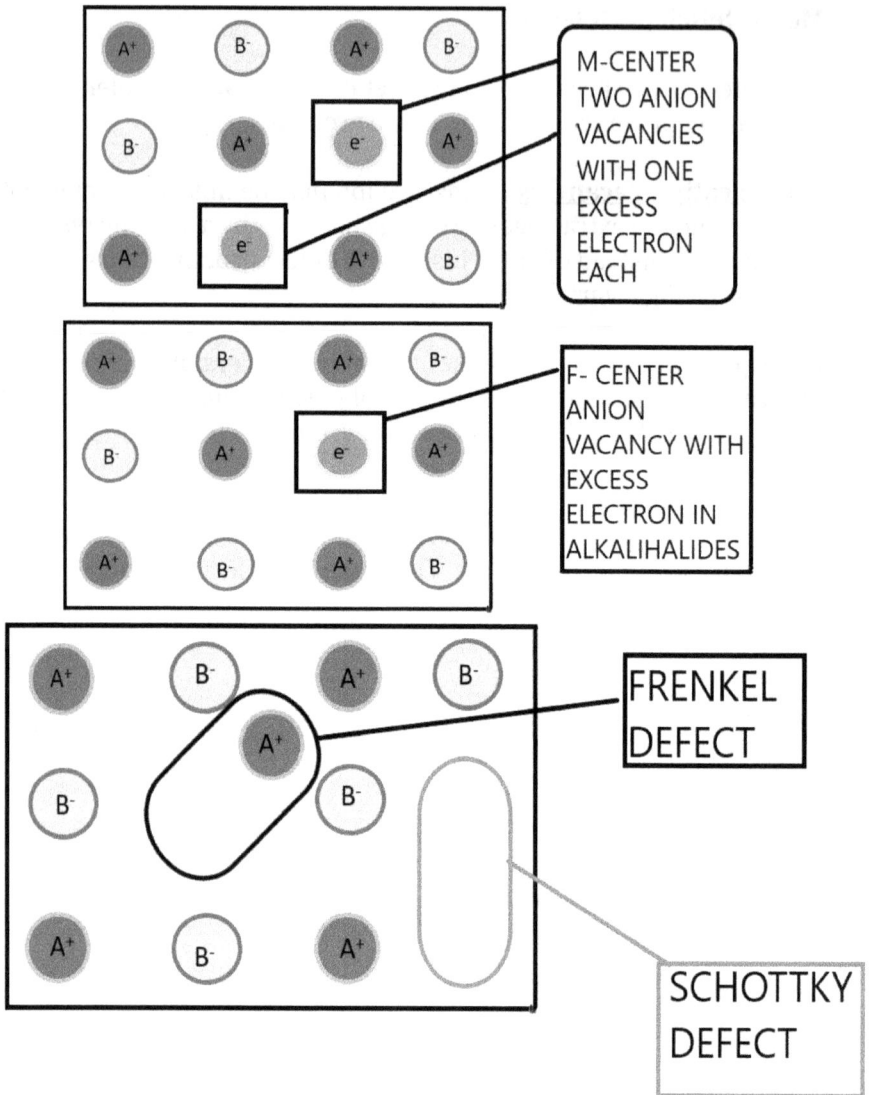

Figure 2.8 Summary of crystal defects.

References

1. Gutierrez-Campos, D., and Diaz, J. I. (1999) Evolution of an alumina-magnesia self forming spinel castable. *Ceramica*, **45**, 292-293.
2. Papike, J. J., Karner, J. M., and Shearer, C. K. (2005) Comparative planetary minerology: Valence state partitioning of Cr, Fe, Ti and V

among crystallographic sites in spinel. *American Mineralogist*, **90**, 277-290.

3. Bosi, F. (2019) Chemical and structural variability in cubic spinel oxides. *Acta Crystallographica*, **B75**, 279-285.

4. Mohan, S. K. (2016) Densification and Characterization of Magnesium Aluminate Spinel from Commercial Grade Reactants: Effect of Milling and Additives. Dissertation NIT Rourkela, India.

5. West, A. R. (2008) *Solid-State Chemistry and its Applications*, Wiley, USA.

6. Fan, H., Yang, Y., and Zacharias, M. (2009) ZnO-based ternary compound nanotubes and nanowires. *Journal of Materials Chemistry*, **19**, 10.

7. Hoque, S. M., Hossain, Md. S., Choudhary, S., Akhter, S., and Hyder, F. (2016) Synthesis and characterization of $ZnFe_2O_4$ nanoparticles and its biomedical applications. *Materials Letters*, **16**(2), 60-63.

8. Martin M. (2002) Life and achievements of Carl Wagner, 100th birthday. *Solid-State Ionics*, **152-153**, 15-17.

9. Nakazima H. (1997) The discovery and acceptance of the KirKendall Effect: The result of a short research career. *Journal of Materials*, **49**(5), 15-19.

10. Gunter, J. R., and Oswal, H.-R. (1975) Attempt to a systematic classification of topotactic reactions. *Bulletin of the Institute for Chemical Research*, **53**(2), 249-255.

11. Barr, L. W., and Lidiard, A. B. (1970) *Defects in Ionic Crystals in Physical Chemistry* (Ed. W. Jost), 10th edition, Academic Press, USA.

12. Fine, M. E. (1973) *Introduction to Chemical and Structural Defects in Crystalline Solids in Treatise on Solid State Chemistry* (Ed. N. B. Hannay), 1st volume, Plenum Press, USA.

13. Kofstad, P. (1972) *Non-Stoichiometry, Electrical Conductivity and Diffusion in Binary Metal Oxides*, Wiley, USA.

14. Mandelcorn, E. (1964) *Non-Stoichiometric Compounds*, Academic Press, USA.

Chapter 3

Experimentation and Kinetics of Solid-State Reactions

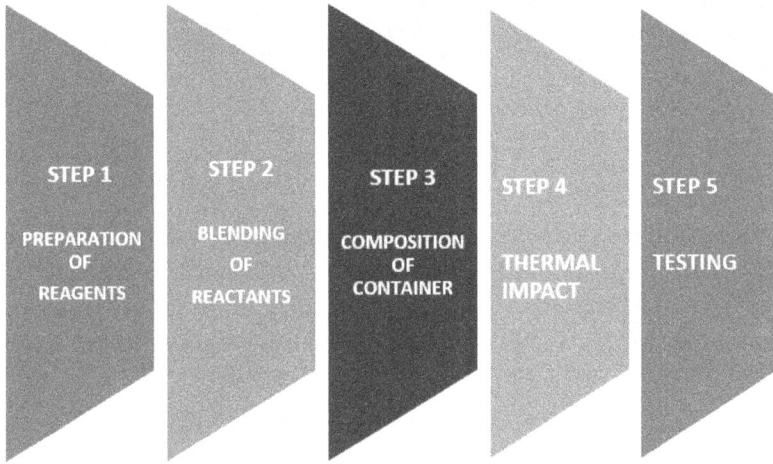

Figure 3.1 Experimental steps involved in solid-state reactions.

Solid state reactions are conducted experimentally. There is a typical experimental procedure proposed for the synthesis of spinel.

3.1 Preparation of Reactants

The reactants are well dried at high temperature for some time duration and weighed. The reagents used for such reactions should be finely grinded to increase the surface area for better reaction rate. The choice of reactants also depends on the nature of compound, if the compound acting as reactant is not able to withhold the experimental conditions, then it may be replaced by other alternatives.

Figure 3.2 Preparation of reactants.

For example, the synthesis of spinel $MgAl_2O_4$, begins with MgO and Al_2O_3 as starting materials. Since magnesium oxide is hygroscopic hence, it may be replaced by magnesium carbonate or any other oxy salts of magnesium. When oxy salt is used as a source of magnesium oxide then conversion becomes the first step followed by the drying weighing and grinding.

3.2 Blending of Reactants

Once the reactants have been prepared and weighed, they are mingled together. The apparatus used for mixing should be unreactive towards the reactants. It should also not contaminate the reactants during the process of mixing. The surface of the apparatus must be smooth and non-porous to facilitate the process of mixing.

1. If the quantity of prepared reactants is small, then simple apparatus is used for mixing. For example, **agate mortar and pestle. Agate** is a common rock formation, consisting of a cryptocrystalline form of silica (chalcedony) and quartz as its core components. Agates are primarily formed within volcanic and metamorphic rocks. Agate is referred in industry due to its hardness, ability to retain a highly polished surface finish and resistance to chemical attack.

Figure 3.3 Mortar and pestle.

For better mixing, volatile organic liquids are added to form a paste. The paste helps in homogenization of the reactants and during the process of blending, the liquid evaporates due to its volatile nature.

2. If the quantity of prepared reactants is large, then small sets of the reactants are prepared and then grinded manually or blending is done mechanically using a **ball mill**.

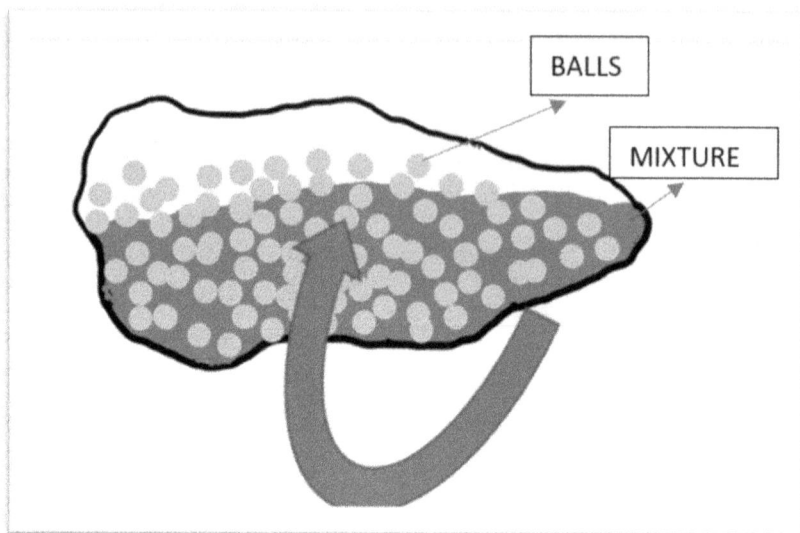

Figure 3.4 Working principle (rotatory) of a ball mill (inside view).

Ball mill is a type of grinder, which is a cylindrical device used in grinding materials. It rotates around a horizontal axis which may be either horizontal or at a small angle to the horizontal. The inner surface of the cylindrical shell is usually lined with an abrasion-resistant material such as manganese steel or rubber lining. The mill is filled with the material to be grinded and the grinding medium. Ceramic balls, flint pebbles, and stainless-steel balls may be used as medium. The grinding works on the principle of critical speed. As the shell rotates, the balls are lifted up on the rising side of the shell and then they cascade down (or drop down on to the feed), from near the top of the shell. The size reduction of the material is done by impact as the balls drop from near the top of the shell.

3.3 Composition of Container

As the solid state reactions occur at high temperature, hence the container in which these reactions are to be conducted have to be chemically inert at such range of temperatures. However, noble metals are quite expensive, yet they are preferred for such reactions. For example, platinum, gold or their alloys.

Crucibles made of alloys tend to be very sturdy and facilitate repeated usage. Reactions of low temperature are conducted in nickel crucible. Many inorganic compounds are inert and can well be used as containers in conducting solid state reactions, like alumina, zirconia, etc. These compounds make the process very cost effective but cannot with stand high temperature.

Alumina crucible   **Zirconia crucible**

Silica crucible

Figure 3.5 Various types of crucibles.

3.4 Thermal Effect

Heat treatment is a particularly important aspect of solid - state reactions. Heat is provided keeping in mind the reactivity and form of the reactants. This is mostly brought about in a furnace, but the reactants must not be directly exposed to extremely high temperature as it may spoil the reaction mixture. For example, Na_2CO_3. Hence, the prepared reactants must be first heated at a lower temperature and then heated at the required temperature. **Tamman's rule** suggests a temperature of about two-thirds of the melting point (K) of the lower melting reactant is needed to have reaction to occur in a reasonable time.

Many a times grinding is conducted along with heating. This aids in maintaining a high surface area to speed up the reaction and it also helps in constantly bringing fresh surface in contact for the reaction. Some compounds tend to change their oxidation states at high temperature hence it becomes crucial to manage same atmosphere throughout the reaction. For example, spinel formation involving Fe^{++} need reducing atmosphere to prevent the conversion to Fe^{+++}.

3.5 Testing

The analysis of the finally obtained crystal is done with **X-ray powder diffraction**. This technique helps in identifying the phase present in the crystal. The X-ray powder diffraction method serves to indicate whether the reaction is complete, by indicating that the actual reactants are absent and by confirming the presence on any unwanted side products/intermediates.

It is based on interference (constructive) of monochromatic X-rays (generated by cathode ray tubes and filtered) and spinel sample. In each set of crystal plane, there will be many crystals in appropriate orientation resulting in correct Bragg angle for Bragg's equation. Every diffraction line constitutes numerous small spots from individual crystal. The spots being infinitesimally small together appear like a continuous line; every crystal plane has a capability of diffraction.

When monochromatic beam of X-ray falls on powdered spinel sample, sharp lines are obtained on the circular photographic film surrounding the sample. On rotation of sample and detector, intensity of reflected X-rays is recorded. When the geometry of the incident X-rays satisfies the Bragg's equation, constructive interface occurs and a peak in intensity occurs. The detector records and processes this X-ray signal and converts it into a count rate which is sent to any output device (like, monitor or printer) for display.

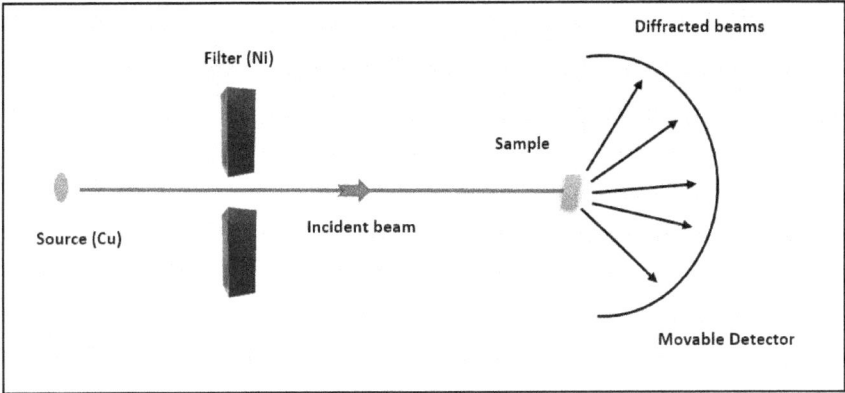

Figure 3.6 Principle of diffraction.

The X-ray powder diffraction does not provide any chemical analysis, however, the **X-ray fluorescence** serves as a good option. It is a chemical analysis method used to determine composition of the sample. The technique is based on the principle that on excitation by external energy source, the atoms emit X-ray photons of a particular energy/wavelength. An account of the number of photons of each energy emitted by the sample the constituent elements in the sample are identified.

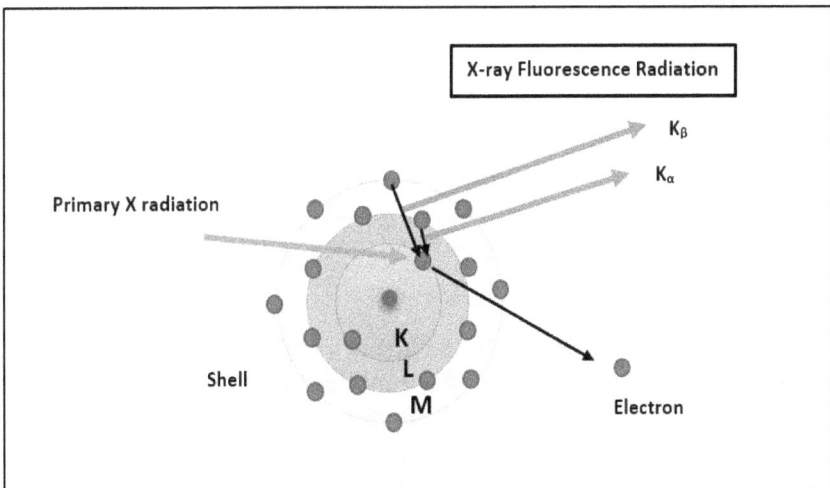

Figure 3.7 Working principle of X-ray fluorescence.

Along with X-ray fluorescence, **atomic absorption spectroscopy** is also used to chemically analyze the spinel to accesses its composition. This method is based on the principle that free atoms produced in an atomizer can absorb radiation at specific frequency and move to higher electronic energy levels. The amount of absorption helps in determining the sample.

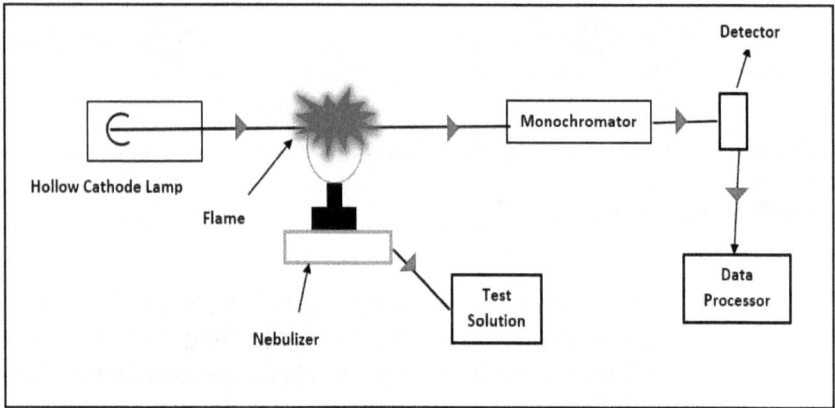

Figure 3.8 Principle of atomic absorption spectroscopy.

3.6 Kinetics of Solid-State Reactions

To understand the kinetics of solid-state reactions, it is important to know the slow and rate determining step of the reaction. The solid-state reaction has three reaction sites:

1. Movement of ions to reaction site
2. Reaction at boundary of contact
3. Movement of products away from reaction boundary

The geometric models of solid-state reactions are based upon the processes of **nucleation** and growth of product nuclei by interface advance. Nuclei are generated predominantly, perhaps exclusively, at or near crystal surfaces and the numbers participating at different times have been measured for some decompositions using microscopy. Several models for nucleus generation are single-step nucleation, instantaneous nucleation, linear nucleation, multi-step nucleation, and branching nucleation. In decompositions of solids where the initiation of reaction results in the rapid development of many closely spaced growth nuclei on all surfaces or on specific crystallographic faces, the overall kinetics of reaction are determined by the geometry of advance of a coherent reaction interface from these boundaries towards the centers of the reactant particles. In reactions of this type, the induction period, if any, may be too short to permit detection, and during this time, there is virtually instantaneous and dense nucleation across all active surfaces.

References

1. Yifeng, W., and Enrique, M. (1990) Self-organizational origin of agates: Banding, fibre twisting, composition and dynamic crystallization model. *Geochimica et Cosmochimica Acta*, **54**(6), 1627-1638.
2. Laszlo, T. (2002) Self-sustaining reactions induced by ball milling. *Progress in Material Science*, **47**(4), 355-414.
3. Florez-Zamora, M. I. (2008) Comparative study of Al-Ni-Mo alloys obtained by mechanical alloying in different ball mills. *Reviews on Advanced Materials Science*, **18**, 301.
4. Majeed, S., and Shivashankar, S. A. (2014) Rapid microwave-assisted synthesis of Gd_2O_3 and Eu: Gd_2O_3 nanocrystals: characterization, magnetic, optical and biological studies. *Journals of Material Chemistry*, **34**.
5. Easwaradas, K. G. (2019) Synthesis of NIR emitting rare earth doped fluorapetite nanoparticles for bioimaging applications. *Current Physical Chemistry*, **9**(2).
6. Andrew K. G., and Michael E. B. (1999) Kinetic model for solid state reactions. *Studies in Physical and Theoretical Chemistry*, Elsevier, **86**, 75-115.
7. Van der Ven, A., Thomas, J. C., Xu, Q., and Bhattacharya, J. (2010) Linking the electronic structure of solids to their thermodynamic and kinetic properties. *Mathematics and Computers in Simulation*, **80**, 1393–1410.

Chapter 4

Structure of Spinels

Introduction

The **spinels** having one atom of A, two atoms of B and four atoms of X comprise of the general chemical formula **AB₂X₄**,

In which:
A^{++} = a divalent cation (Mg, Cr, Mn, Fe, Co, Ni, Cu, Zn, Cd, Sn)
B^{+++} = a trivalent cation (Al, Ga, In, Ti, V, Cr, Mn, Fe, Fe, Co, Ni)
X^{--} = O, S, Se etc.

Such type of spinels are normal spinels (already discussed in **Chapter 2**).

4.1 Crystal Structure of Normal Spinels

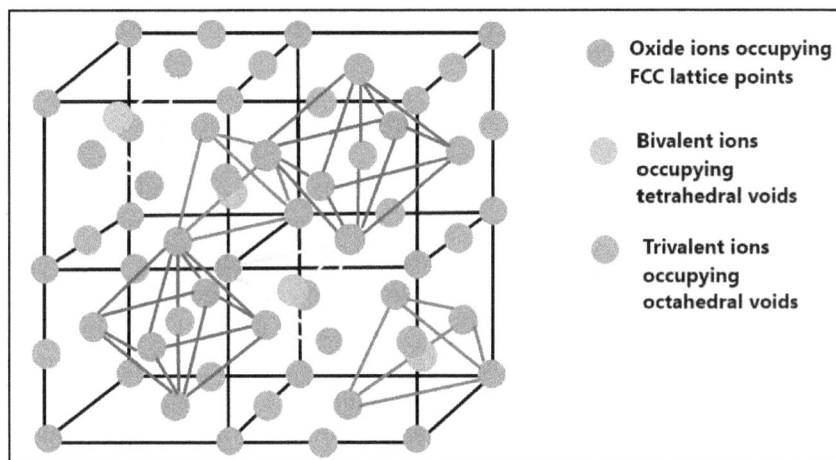

There are **eight face centric cubic** (FCC) cells in one unit cell of normal spinel (**AB₂O₄**). The FCC lattice points are occupied by the negatively charged ions (i.e., anions usually oxide ions: O^{2-}). One-eighth of the tetrahedral voids are inhabited by the divalent cations (A^{II}) and

one-half of octahedral voids are occupied by the trivalent cations (B^{III}). Now, in **one FCC lattice unit cell**, the effective number of atoms (or anions) occupying the lattice points is 4. Accordingly, the effective number of tetrahedral voids (holes) = 8 and the effective number of octahedral voids = 4. Hence, in a normal spinel there are 8 FCC cells,

1. The number of tetrahedral voids (occupied by divalent A^{II} cations) = **8** (8 x 1/8 x 8)

2. The number of octahedral voids (occupied by trivalent B^{III} cations) = **16** (8 x 1/2 x 4).

3. The number of anions occupying the lattice points of 8 FCC unit cells = **32** (8 x 4).

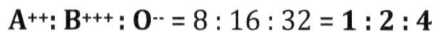

$$A^{++}: B^{+++} : O^{--} = 8 : 16 : 32 = \mathbf{1 : 2 : 4}$$

This ratio is in synchronization with the formula of normal spinel. Thus, a **normal spinel** can be represented as: $(A^{II})^{tet}(B^{III})_2{}^{oct}O_4$.

Examples of normal spinels: $MgAl_2O_4$ (known as spinel), Mn_3O_4, $ZnFe_2O_4$, $FeCr_2O_4$ (chromite) etc.

4.2 Crystal Structure of Inverse Spinels

There are **eight face centric cubic** (FCC) cells in one unit cell of inverse spinel **($B(AB)O_4$))**. The FCC lattice points are occupied by the negatively charged ions (i.e., anions usually oxide ions: O^{2-}). One-fourth of the octahedral voids are inhabited by the divalent cations (A^{II}), and one-half of one-eighth of tetrahedral voids are occupied by the trivalent cations (B^{III}) and other-half of one-fourth of octahedral voids are occupied by the trivalent cations (B^{III}). Now, in **one FCC lattice unit cell**, the effective number of atoms (or anions) occupying the lattice points is 4. Accordingly, the effective number of tetrahedral voids (holes) = 8 and the effective number of octahedral voids = 4. Hence, in an inverse spinel there are 8 FCC cells,

1. The number of **octahedral voids** (occupied by divalent A^{II} cations) = **8** (8 x 1/4 x 4)

28

2. The number of **tetrahedral voids** (occupied by trivalent B^{III} cations) = **8** (8 x 1/8 x 8).

3. The number of octahedral voids (occupied by trivalent B^{III} cations) = **8** (8 x 1/4 x 4).

4. The number of anions occupying the lattice points of 8 FCC unit cells = **32** (8 x 4).

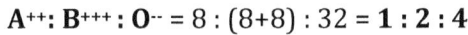

$$A^{++}: B^{+++}: O^{--} = 8 : (8+8) : 32 = \textbf{1 : 2 : 4}$$

This ratio is in synchronization with the formula of inverse spinel. Thus, an **inverse spinel** can be represented as: $(B^{III})^{tet}(A^{II}B^{III})^{oct}O_4$.

Examples of inverse spinels: Fe_3O_4 (ferrite), $CoFe_2O_4$, $NiFe_2O_4$ etc.

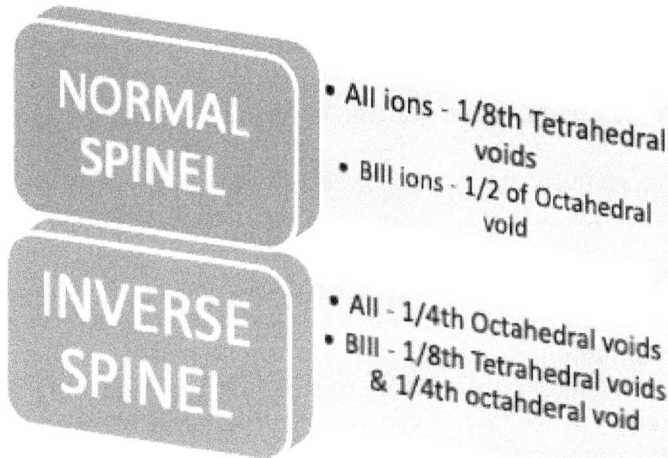

NORMAL SPINEL	• All ions - 1/8th Tetrahedral voids • BIII ions - 1/2 of Octahedral void
INVERSE SPINEL	• All - 1/4th Octahedral voids • BIII - 1/8th Tetrahedral voids & 1/4th octahderal void

4.3 Factors Affecting the Structure of Spinels

A. Size of cation (A and B): Generally, the lower co-ordination site (tetrahedral site) is occupied by the smaller cation. But, in $MgAl_2O_4$ spinel, this point is violated because Al^{3+} being the smaller sized cation should have occupied tetrahedral void, but it occupies octahedral positions. This is so because of higher lattice energy.

B. Ligand-field stabilization energies: Ligand-field stabilization energy (LFSE) is applied for transition metal ions and LFSE has an impact on the structures of spinels. The ion with more LFSE value in octahedral position as compared to the LFSE in tetrahedral positions tends to occupy octahedral sites. This gap between the LFSE values in octahedral and tetrahedral positions is known to as **octahedral site stabilization energy** (OSSE).

i.   If the B^{III} ion has more CFSE (Crystal Field Stabilization Energy) gain in octahedral site than that of A^{II} ion, a **normal spinel** structure is expected.

ii. If A^{II} ion has more CFSE gain in octahedral geometry than the B^{III} ion, an **inverse spinel** is formed.

C. The Madelung constant: The value of Madelung constants for the normal and inverse spinel structures are same and hence these values do not play much role in understanding the type of spinels.

4.4 Rules for Structure Formation of Spinels

In a normal spinel structure, no **crystal field stabilization energy** is involved and the divalent and the trivalent metals are non-transition metals.

1. In some cases, there is a tendency of formation of inverse spinel structure containing transition metal ions. The reason for this is that the transition metal ion gets increased stability (Ligand Field Stabilization Energy) in octahedral geometry, and hence prefers octahedral voids over tetrahedral ones.
2. The ions having d^0; high spin d^5, d^{10} orbitals have no proclivity between tetrahedral and octahedral coordination since their LFSE is zero.
3. Generally, ions having d^3 & d^8 orbitals have strongest preference for octahedral geometry.
4. Other ions having d^1, d^2, d^4, d^6, d^7, d^9 orbitals too have slightly more propensity for octahedral symmetry.

For example, the **spinel is INVERSE, if A^{II} has d^3 or d^8 configuration** and the B^{III} ion has configuration other than these.

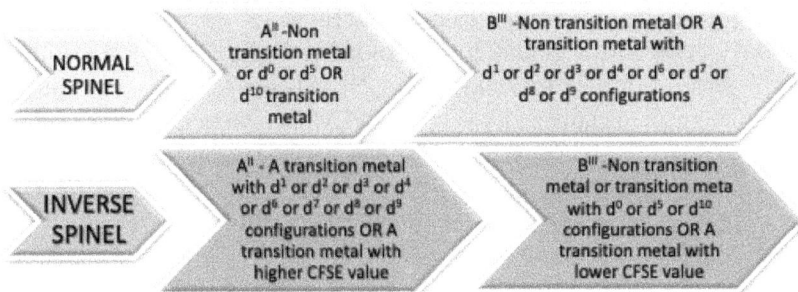

NORMAL SPINEL	A^{II} -Non transition metal or d^0 or d^5 OR d^{10} transition metal	B^{III} -Non transition metal OR A transition metal with d^1 or d^2 or d^3 or d^4 or d^6 or d^7 or d^8 or d^9 configurations
INVERSE SPINEL	A^{II} - A transition metal with d^1 or d^2 or d^3 or d^4 or d^6 or d^7 or d^8 or d^9 configurations OR A transition metal with higher CFSE value	B^{III} -Non transition metal or transition metal with d^0 or d^5 or d^{10} configurations OR A transition metal with lower CFSE value

4.5 Examples of Spinel Structures

1. Since both the divalent (Mg^{++}) and trivalent ions (Al^{+++}) are non-transition metal ions and there is no point of CFSE, hence **$MgAl_2O_4$ has normal spinel** structure.
2. Since the Mn^{++} ion is a high spin d^5 system with zero LFSE and Mn^{+++} ion is a high spin d^4 system with considerable LFSE, hence **Mn_3O_4 is a normal spinel**.
3. Since the Fe^{++} is a high spin d^6 system with more CFSE and Fe^{+++} ion is a high spin d^5 system with zero CFSE, hence **Fe_3O_4 has an inverse spinel structure**.
4. Even in the presence of weak field oxo ligands, the Co^{+++} is a low spin d^6 ion with very high CFSE. It is due to high charge on Co^{+++}. Hence all the Co^{+++} ions occupy the octahedral sites, hence **Co_3O_4 is a normal spinel**.
5. Since the Ni^{++} is a d^8 ion having more CFSE than the Fe^{+++} which is a d^5 ion, hence **$NiFe_2O_4$ is an inverse spinel**.

4.6 Applications of Spinels

1. Spinels have found great usage as **gemstones** due to their greater refractive index. These spinel gemstones are considered very rare and precious. For example, Samarian Spinel is the fourth largest Spinel and is a 500-carat spinel gemstone.

2. **Perovskites solar cells** are a type of solar cell including a perovskite absorber. It acts as the light-harvesting active layer, producing electricity from sunlight. The major benefits of Perovskite absorber materials are that they are cheap to produce and simple to manufacture.
3. Perovskite can also generate **laser light**. Methyl ammonium lead iodide perovskite ($CH_3NH_3PbI_{3}$-xCl_x) cells can be designed into optically pumped VCSELs (vertical-cavity surface-emitting lasers) which convert visible pump light to near-IR laser light efficiently.
4. Spinels are generally used in TV and phone **screens.**
5. Spinels also find use as superconducting magnet in MRI and NMR **supermagnets**.
6. They also register usage in **photo-electrolysis**. Using perovskite photovoltaics, electrolysis of water can be efficiently done in a highly coherent and low-cost water-splitting cell.

References

1. Van der Ven, A., and Ceder, G. (2005) Vacancies in ordered and disordered binary alloys treated with the cluster expansion. *Physical Reviews B*, **71**, 94–102.
2. Wood, B. J., Kirkpatrick, R. J., and Montez, B. (1986) Order-disorder phenomena in $MgAl_2O_4$ spinel. *American Mineralogist*, **71**, 999–1006.
3. Andreozzi, G. B., Princivalle, F., Skogby, H., and Della Giusta, A. (2000) Cation ordering and structural variations with temperature in $MgAl_2O_4$ spinel: An X-ray single-crystal study. *American Mineralogist*, **85**, 1164–1171.
4. Maekawa, H., Kato, S., Kawamura, K., and Yokokawa, T. (1997) Cation mixing in natural $MgAl_2O_4$ spinel: A high-temperature 27Al NMR study. *American Mineralogist*, **82**, 1125–1132.
5. Schmocker, U., Boesch, H. R., and Waldner, F. (1972) A direct determination of cation disorder in $MgAl_2O_4$ spinel by ESR. *Physics Letters A*, **40**, 237–238.
6. Cynn, H., Anderson, O. L., and Nicol, M. (1993) Effects of cation disordering in a natural $MgAl_2O_4$ spinel observed by rectangular parallelepiped ultrasonic resonance and Raman measurements. *Pure and Applied Geophysics*, **141**, 415–444.
7. De Fontaine, D. (1994) *Solid State Physics*, volume 47, Elsevier, USA.
8. Redfern, S. A., Harrison, R. J., O'Neill, H. S. C., and Wood, D. R. (1999) Thermodynamics and kinetics of cation ordering in $MgAl_2O_4$ spinel

up to 1600 °C from in situ neutron diffraction. *American Mineralogist*, **84**, 299–310.

9. Van Minh, N., and Yang, I.-S. (2004) A Raman study of cation-disorder transition temperature of natural $MgAl_2O_4$ spinel. *Vibrational Spectroscopy*, **35**, 93–96.

10. Seko, A. (2006) First-principles study of cation disordering in $MgAl_2O_4$ spinel with cluster expansion and Monte Carlo simulation. *Physical Reviews B*, **73**, 94-116.

11. Hill, R. J., Craig, J. R., and Gibbs, G. V. (1979) Systematics of the spinel structure type. *Physics and Chemistry of Minerals*, **4**, 317–339.

12. Liu, J. (2019) Unified view of the local cation-ordered state in inverse spinel oxides. *Inorganic Chemistry*, **58**, 14389–14402.

13. Jiang, C., Sickafus, K. E., Stanek, C. R., Rudin, S. P., and Uberuaga, B. P. (2012) Cation disorder in MgX_2O_4 (X= Al, Ga, In) spinels from first principles. *Physical Reviews B*, **86**, 94-203.

14. Molla, A. R. (2014) Microstructure, mechanical, thermal, EPR, and optical properties of $MgAl_2O_4$:Cr_3 spinel glass-ceramic nanocomposites. *Journal of Alloys and Compounds*, **583**, 498–509.

Chapter 5

Co-precipitation and Other Precursor Techniques

The traditional method for solid state reactions has many drawbacks because of which **precursor techniques** are employed. These methods use precursors such as nitrates and carbonates as starting materials instead of oxides. In precursor method, atomic level mixing occurs by forming a solid compound (precursor) containing the metals of desired compound in proper stoichiometry. Products from precursor methods often contain small particles with a large surface area, which is desired for certain applications.

Co-precipitation procedures helps in achieving extreme degree of homogenization with small size of particle thereby speeding the reaction. It involves taking a stoichiometric mixture of soluble salts of the metal and precipitating them as hydroxides, citrates, oxalates or formates. The mixture is filtered, dried, and then heated to give the final product. Some of the methods are listed below:

5.1 Sol-gel Precipitation Method

Sol-gel precipitation method has multiple steps for the formation of solid-state products. The precipitation method is initiated by the formation of sol by hydrolysis if metal organic reactant in organic solvent which is miscible with water/aqueous inorganic salt solution. This step is followed by the conversion of sol into gel by condensation and polycondensation. Finally, evaporation and sintering yields nanocrystalline powder.

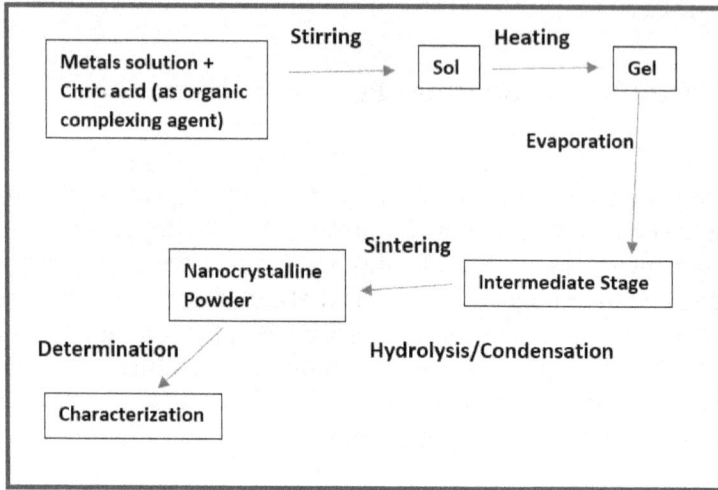

Figure 5.1 Sol-gel precipitation method.

Another example is formation of **lithium niobate, LiNbO₃**. The formation starts with lithium ethoxide ($LiOC_2H_5$) and niobium ethoxide [$Nb_2(OC_2H_5)_{10}$] dissolved in absolute EtOH and mixed. Addition of water leads to partial hydrolysis giving hydroxyalkoxides. Hydroxyalkoxides condense to form a polymeric gel with metaloxygen-metal links. $LiNbO_3$ is formed when the gel is heated – the remaining ethanol and water is evaporated, and any remaining ethyl groups are pyrolyzed (forming CO_2 and H_2O).

5.2 Microwave Method

Microwave method applies if no particles are present that can move freely, but molecules or units with dipole moments are present, then the electric field acts to align the dipole moments. This is dielectric heating. This is the type of heating that acts on water molecules in food. The electric field of the microwave radiation is oscillating at the frequency of the radiation, but the electric dipoles in solids do not change their alignment instantaneously, but with a characteristic time, t. The oscillating electric field changes its direction rapidly so that the time between changes is much smaller than t, then the dipoles cannot respond fast enough and do not realign. The solid absorbs some of the microwave radiation and the energy is converted to heat.

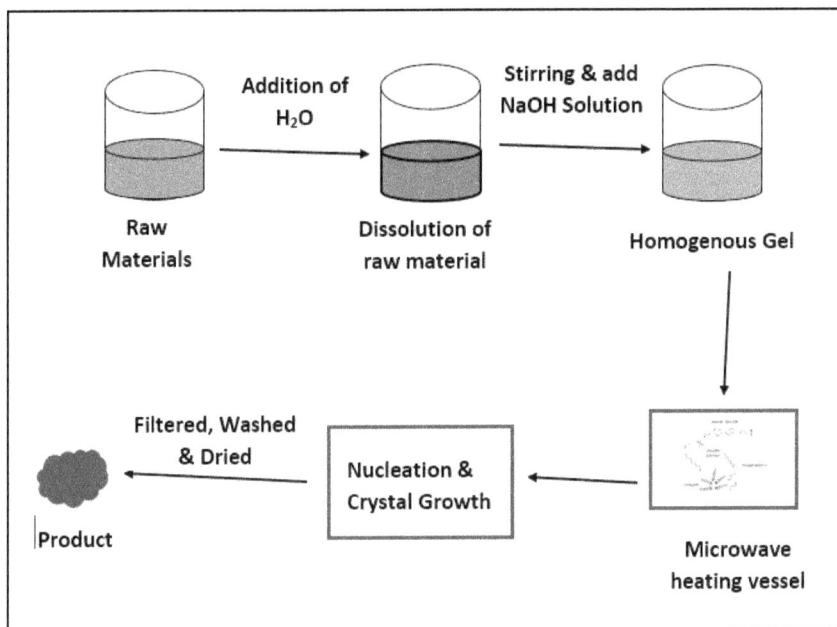

Figure 5.2 The microwave method.

5.3 High Pressure Method/Hydrothermal Method

High pressure method or **hydrothermal method** involves heating of reactants in water/steam at high pressures and temperatures. Water, in this method, executes two functions; as pressure transmitting medium and as a solvent, in which the solubility of the reactants is pressure-temperature dependent. Reactants and water are placed inside a "**BOMB**" which is either sealed or connected to an external pressure control. This is placed in an oven with a temperature range of 100-500°C.

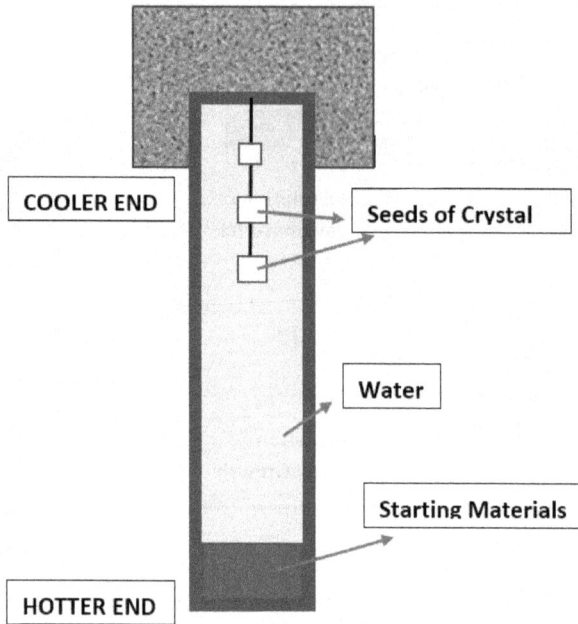

Figure 5.3 High pressure/hydrothermal method.

An example of hydrothermal method is synthesis of **calcium silicate hydrates** which is prepared from lime and quartz on heating (Temperature = 150 - 500°C) in water under pressure (0.1 - 2.0 kbar).

5.4 Czochralski Process

This method helps in growth of the crystal from the molten matter(melt) of same composition where a seed crystal is placed in contact of the molten matter. The temperature of the melt is kept slightly above its melting point. As the shaft is lifted with the seed, the molten matter solidifies on the surface of the of the seed. The uniformity in temperature and coating is maintained by rotating the shaft clockwise and anticlockwise. For example, semiconducting material, Si and Ge are widely prepared by this method.

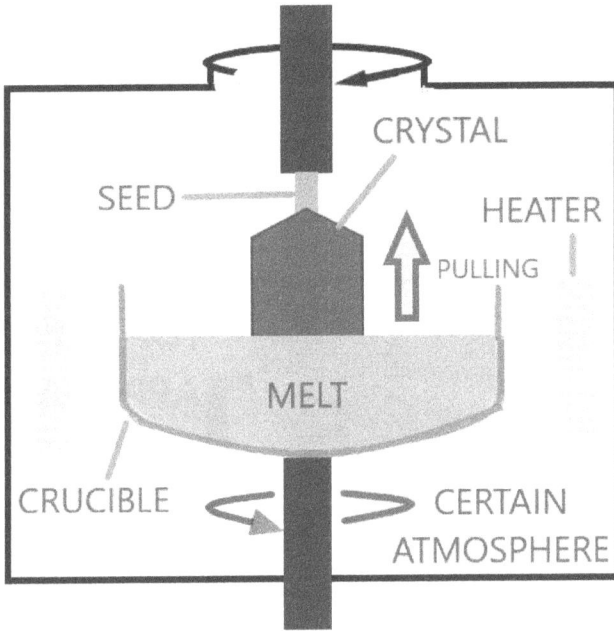

Figure 5.4 The Czochralski process.

5.5 Top-Down and Bottom-Up Methods

Top-down approach refers to slicing or successive cutting of a bulk material to get nano sized particle. Bottom-up approach refers to the buildup of a material from the bottom: atom by atom, molecule by molecule or cluster by cluster. Both approaches play important role in modern industry and most likely in solid-state reactions as well. There are advantages and disadvantages in both approaches.

Attrition or milling is a typical top-down method in making nano particles, whereas the colloidal dispersion is a good example of bottom-up approach in the synthesis of nano particles. The biggest problem with top-down approach is the imperfection of surface structure and significant crystallographic damage to the processed patterns. These imperfections which in turn leads to extra challenges in the device design and fabrication. But this approach leads to the bulk production of material. Regardless of the defects produced by top-down approach, they will continue to play an important role in the synthesis of structures.

Though the bottom-up approach often referred in solid-state technology, it is not a newer concept. All the living beings in nature observe growth by this approach only and it has been in industrial use for over a century. Examples include the production of salt and nitrate in chemical industry. Although the bottom-up approach is nothing new, it plays an important role in the fabrication and processing of structures.

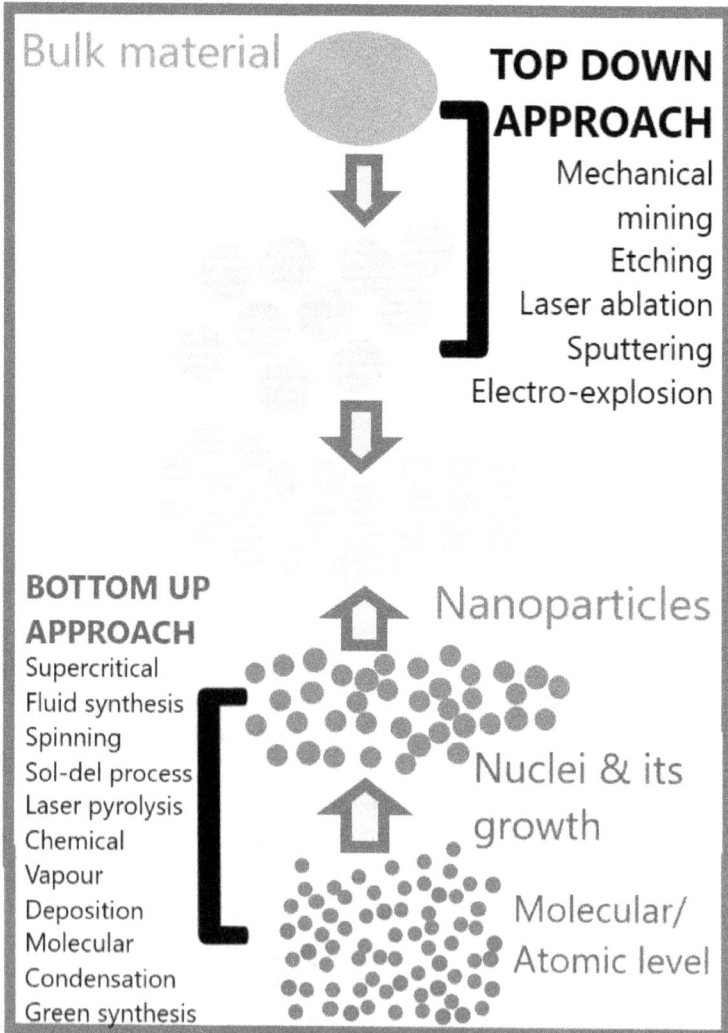

Figure 5.5 Top-down and bottom-up methods.

References

1. Mirjalili, F., Mohamad, H., and Chuah, L. (2011) Preparation of nano scale α-Al_2O_3 powder by the sol-gel method. *Ceramics – Silikáty*, **55**(4), 378–383.
2. Rogojan, R., Andronescu, E., Ghitulica, C., and Stefan, B. (2011) Synthesis and characterization of alumina nano-powder by sol-gel method. *UPB Scientific Bulletin, Series B*, **73**(2), 67–76.
3. Mishara, D., Anand, S., Panda, R. K., and Das, R. K. (2000) Hydrothermal preparation and characterization of boehmites. *Materials Letters*, **42**, 38-45.
4. Kamata, K., Mochizuki, T., Matsumoto, S., Yamada, A., and Miyokawa, K. (1985) Preparation of submicrometer Al_2O_3 powder by gas-phase oxidation of tris (acetylacetonato) aluminum (111). *Journal of American Ceramic Society*, **68**, 193–194.
5. Müller, D., Schwerin, J., Gille, P., and Fehr, K. (2014) High-resolution EPMA X-ray images of mother liquid inclusions in a Pd_2Ga single crystal. *IOP Conference Series: Materials Science and Engineering*, **55**, 012013. Online: 10.1088/1757-899X/55/1/012013.
6. Jesenovec, J., Varley, J., Karcher, S. E., and McCloy, J. S. (2021) Electronic and optical properties of Zn-doped β-Ga_2O_3 Czochralski single crystals. *Journal of Applied Physics*, **129**, 225702.
7. Sabatier, P. A. (1986) Top-down and bottom-up approaches to implementation research: A critical analysis and suggested synthesis. *Journal of Public Policy*, **6**(1), 21–48.
8. Iqbal, P., Preece, J. A., and Mendes, P. M. (2012) Nanotechnology: The "top-down" and "bottom-pp" approaches. In: *Supramolecular Chemistry*, Wiley, USA.
9. Benjamin P. I., and Keith A. B. (2017) Progress in top-down control of bottom-pp assembly. *Nano Letters*, **17**(11), 6508-6510.
10. Djurišić, A. B., Chen, X., Leung, Y. H., and Ng, A. M. C. (2012) ZnO nanostructures: growth, properties and applications. *Journal of Materials Chemistry*, **22**(14), 6526–6535.
11. Gentile, A., Ruffino, F., and Grimaldi, M. G. (2016) Complex-morphology metal-based nanostructures: fabrication, characterization, and applications. *Nanomaterials*, **6**(6), 110.
12. Wang, Y., Iqbal, Z., and and Mitra, S. (2005) Microwave-induced rapid chemical functionalization of single-walled carbon nanotubes. *Carbon*, **43**(5), 1015–1020.
13. Nüchter, M., Ondruschka, B., Bonrath, W., and Gum, A. (2004) Microwave assisted synthesis - a critical technology overview. *Green Chemistry*, **6**, 128–141.
14. Lidström, P., Tierney, J., Wathey, B., and Westman, J. (2001) Microwave assisted organic synthesis - a review. *Tetrahedron*, **57**(51), 9225–9283.

15. Das, S., Mukhopadhyay, A. K., Datta, S., and Basu, D. (2009) Prospects of microwave processing: an overview. *Bulletin of Materials Science*, **32**(1), 1–13.
16. Gedye, R. N., Smith, F. E., and Westaway, K. C. (1988) The rapid synthesis of organic compounds in microwave ovens. *Canadian Journal of Chemistry*, **66**(1), 17–26.
17. Priyadarshana, G., Kottegoda, N., Senaratne, A., Alwis, A. de, and Karunaratne, V. (2015) Synthesis of magnetite nanoparticles by top-down approach from a high purity ore. *Journal of Nanomaterials*, **1**(1), 1-8.

Chapter 6

Organic Metals and Electrical Conductivity in Organic Materials

Introduction

Organic metals comprise of metals (ionic part) along with ligands (organic part) which are connected (**conducting polymers**). The widespread attention on **conducting polymers** is due to its extraordinary properties such as simple preparation step, low cost of monomers, environmentally friendly, and having high conducting properties like metals. In addition, these polymers are lightweight and non-corrosive in nature, which has made them one of the versatile polymers in the materials group. These materials have clusters or single metal ions coordinately attached to a ligand that constitutes the organic site forming single/multi-dimensional structure. Thus, metallic part acts as a node and organic part as a linkage.

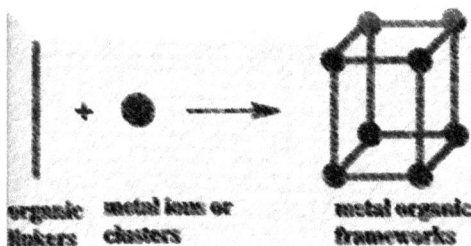

organic linkers metal ions or clusters   metal organic frameworks

Thus, organic metal is a material which is carbon-based (organic) but has properties of a metal, like conductivity, that cannot be expected from an organic substance. It is amazing how organic metals have special properties that make them useful for corrosion protection, oxidation prevention, final finishes and even as catalysts. They have many advantages as being environmentally benign, requiring less energy to apply, reducing raw material usage and are generally non-hazardous.

The organic solids do not conduct electricity since there are no free electrons present in the molecule to carry energy but when double and single bonds are present in alternate arrangement i.e., conjugation, the conductivity values are reported. For example, poly-acetylene has a medium electrical conductivity of 10^{-9} ohm^{-1}cm^{-1} – 10^{-5} ohm^{-1}cm^{-1}.

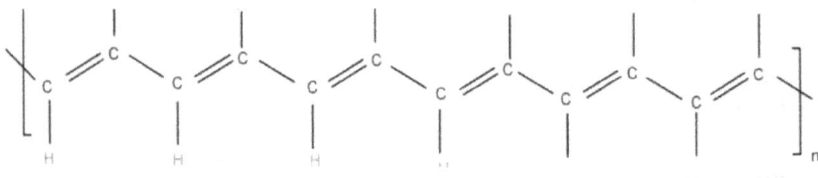

However, if **polyacetylene** is treated with appropriate dopants (like, electron acceptors as bromine, sulfuric acid, and electron donors as alkali metals), its conductivity increases significantly (10^3 ohm^{-1}cm^{-1}). Since their conductivity is nearing metals hence, they are termed as **synthetic or organic metals**.

6.1 Conducting Polymers

Conducting polymers are not thermoplastics and have high electrical conductivity, but they have different mechanical properties as compared to other polymers. These conducting polymers carbon centers which are sp^2 hybridized. Each center has one valence electron in p-orbital which is orthogonal to other three sigma bonds.

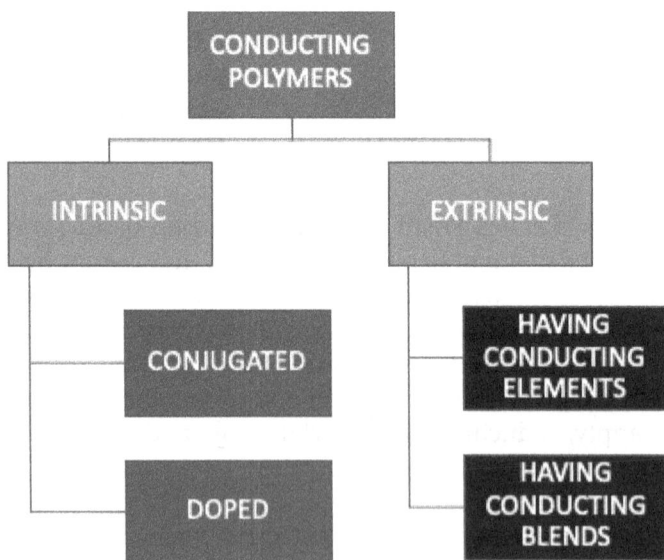

6.2 Intrinsically Conducting Polymers

Intrinsically conducting polymers having **sigma-pi conjugation** exhibit conduction of electricity due to double bonds and lone pair of electrons. The overlapping of conjugated pi-electrons develops valence and conduction bands through the complete structure of the polymer. Conduction of electricity can only occur when the required energy of activation is attained, either by heat or by light.

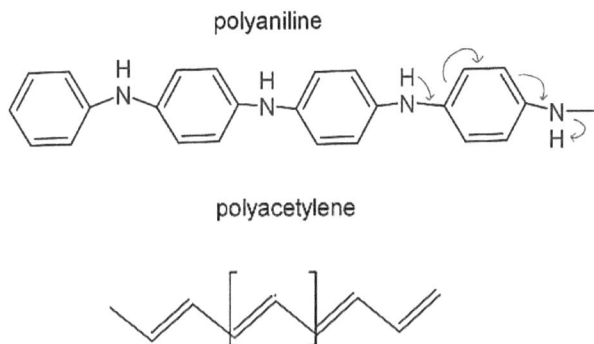

polyaniline

polyacetylene

Intrinsically conducting polymers obtained by **doping** can be of two types: p-type doped conducting polymers and n-type doped conducting polymers. **P-type doped conducting polymers** have conjugation in their structure.

Types of doping
1. **OXIDATION WITH HALOGEN** – Polymer + X = (polymer)$^{n+}$ + X^{n-}
2. **REDUCTION WITH ALKALI METAL** - Polymer + M = (polymer)$^{n-}$ + M^{n+}
X = I$_2$, Br$_2$, AsF$_5$, ...   M = Na, Li, ...

On treatment with Lewis acids i.e., electron-deficient varieties (for example, FeCl$_3$ or I$_2$ vapors or I$_2$-CCl$_4$ couple) oxidation occurs creating positive charge in the molecule. Removal of an electron from the pi-backbone structure of a conjugated polymer forms a radical cation (called **polaron**), which on losing another electron gives **bipolaron.**

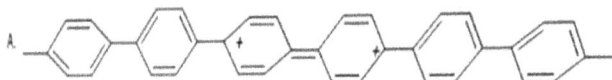

Polaron **(A)**

45

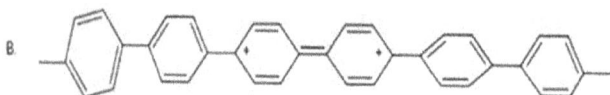

B.

Bipolaron (**B**)

There occurs delocalization of positive charges resulting in Conduction of Electricity.

Polymer+Lewis Acid $\longrightarrow$ p-Doped polymer (oxidative coupling)

$(CH)_x$ + $3I_2$ $\longrightarrow$ $2(CH)_x^+ I_3^-$

(Polyacetylene)

N-type doped conducting polymers on treatment with Lewis bases i.e., electron-rich varieties reduction occurs creating negative charge in the molecule. On addition of one electron, polaron is formed which becomes a bipolaron when the second electron adds and due to this delocalized charge, electrical conduction takes place.

Polymer+ Lewis Base $\longrightarrow$ n- Doped polymer (Reductive coupling)

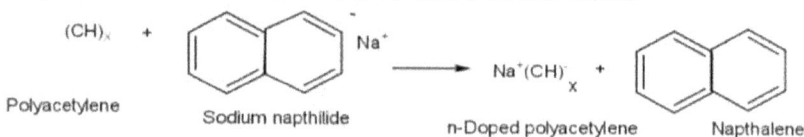

$(CH)_x$ + [Sodium napthilide] Na^+ $\longrightarrow$ $Na^+(CH)_x^-$ + [Napthalene]

Polyacetylene   Sodium napthilide   n-Doped polyacetylene   Napthalene

The intrinsically conducting polymers, along with electrical conductivity, can also store charge, have the capability of ion-exchange, can give coloured compounds by absorbing radiation of visible light.

Plot of conductivity vs doping

Conductivity increases upto a certain doping level

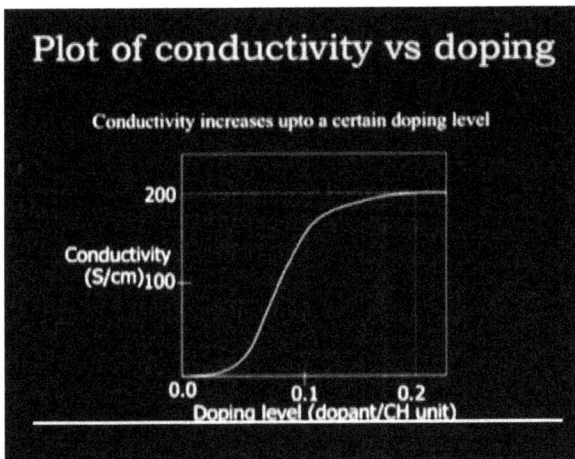

6.3 Extrinsically Conducting Polymers

Extrinsically conducting polymers exhibit conducting properties due to addition of external agents. The minimum amount of additive required to induce conductivity in the compound is called Percolation Threshold. The additive should possess certain traits like it should have more surface area, it should have a filamentous nature and there should be adequate porosity in the material for conducting properties.

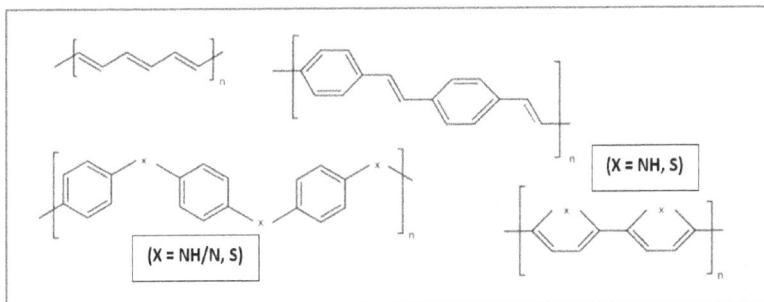

The preparation of such materials is achieved by first preparing the organic part and then doping it with liquid or gaseous dopant. As can be seen in the case of polyacetylene, acetylene is produced by any of three methods: by reaction of water with calcium carbide, by passage of a hydrocarbon through an electric arc, or by partial combustion of methane with air or oxygen. Once acetylene is prepared, at room

47

temperature, a mixture of cis and trans polyacetylene is obtained. Lower temperatures generally produce cis – forms while higher temperatures give trans – forms as products. Trans–polyacetylene is also found to be more stabler than cis – polyacetylene.

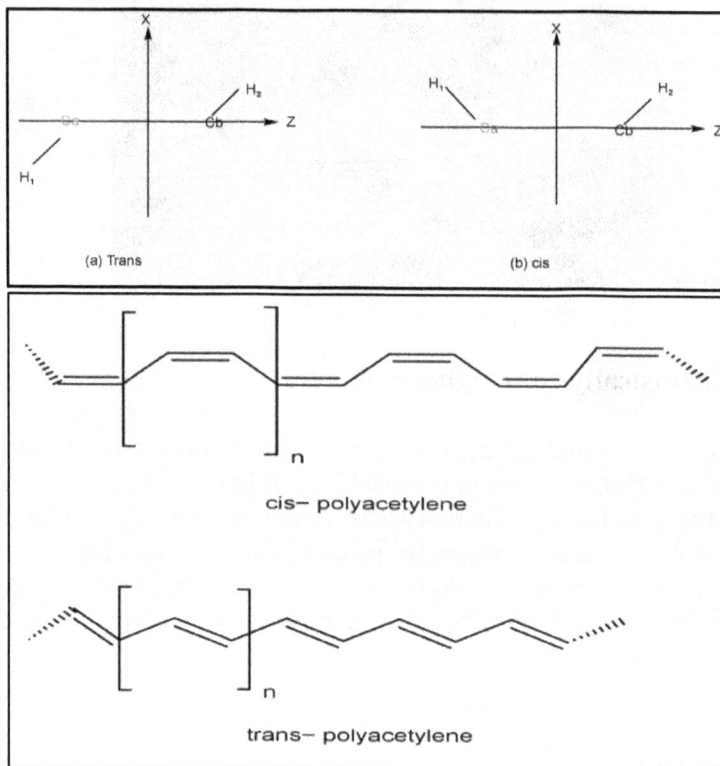

(a) Trans   (b) cis

cis– polyacetylene

trans– polyacetylene

Another example is of **polyaniline** which is a nitrogen hetero-atom aromatic conducting polymer. As a bulk conducting plastic, it is generally not useful. However, the solution of polyaniline proves useful in various applications. Therefore, it is dispersed in various solvents, including aqueous acid solutions, where we can now take advantage of its special properties and environmental benefits. This ability to create stable solvent dispersions of polyaniline results in **"organic metal"**.

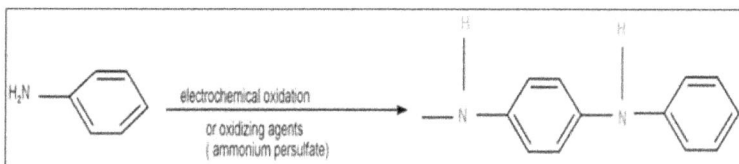

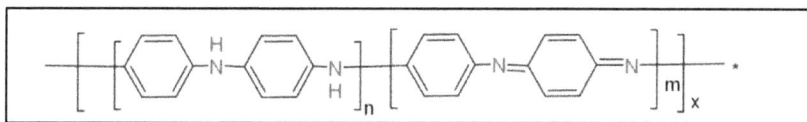

In this figure, n + m =1 and x=degree of polymerization.

Material	Conductivity (S cm⁻¹)
Common Polymer	$<10^{-9}$
Organic Metal	~ 1.0
Typical Metal	$>10^4$

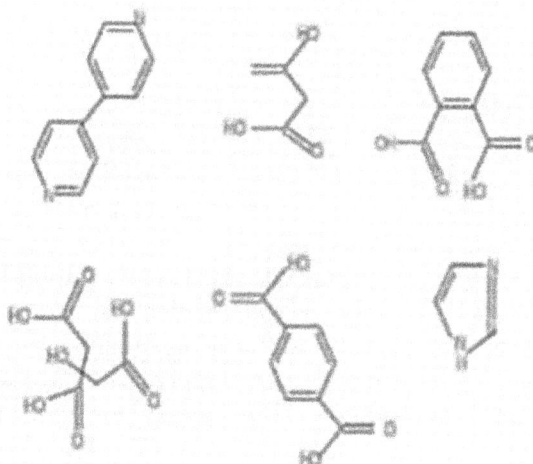

Other organic ligands

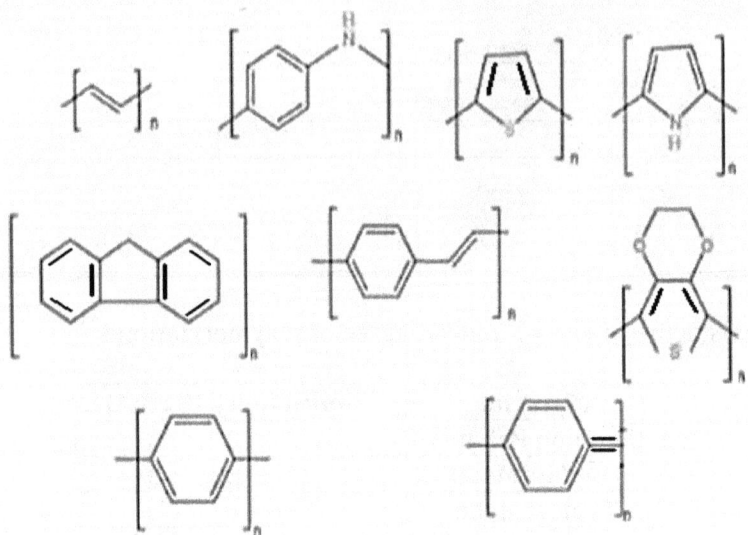

Other organic ligands

6.4 Applications

Conducting polymers/ organic metals are used as **coatings** as they prevent buildup of static charge in insulators, absorb harmful radiations from electrical appliances that are harmful to the nearby gadgets and they are used in circuit boards. These compounds find good usage as **sensors** as in poly-pyrroles which are used to detect NO_2 and NH_3 gases by changing conductivity; as **biosensors** like polymerization of polyacetylene in presence of glucose oxidase (enzyme) and triiodide (redox mediator) to produce a polymer which acts as a glucose sensor.

As the conducting polymers are light-weight and rechargeable hence they find good usage as **batteries**, for example, polypyrrole-Li and polyanniline-Li batteries. Also, these compounds can be used as **conductive Adhesives** in a way that monomers are placed between two conducting plates, sticking them together yet allowing electric current to pass through the bonds.

References

1. Rao, P., and Geckeler, K.E. (2011) Polymer nanoparticles: preparation techniques and size-control parameters. *Progress in Polymer Science*, **36,** 887-913.
2. Bagdžiūnas, G. (2020) Theoretical design of molecularly imprinted polymers based on polyaniline and polypyrrole for detection of tryptophan. *Molecular Systems Design & Engineering*, **5**(9), 1504-1512.
3. Ting, M., Narasimhan, B., Travas-Sejdic, J., and Malmstrom, J. (2021) Soft conducting polymer polypyrrole actuation based on poly(N-isopropylacrylamide) hydrogels. *Sensors and Actuators, B: Chemical*, **343**, 130167.
4. Bredas, J. L., Chance, R. R., and Silbey, R (1982) Comparative theoretical study of the doping conjugated polymers; Polarons in polyacetylene & polyparaphenylene. *Physical Review B*, **26**(10), 5843 - 5854.
5. Clarke, T. C., Geiss, R. H., Kwak, J. F., and Street, G. B. (1978) Highly conducting transition metal derivatives of Polyacetylene. *Journal of the Chemical Society, Chemical Communications*, **338**, 489-490.
6. Yang, C. Y., Cao, Y., Smith, P., and Heeger, A. J. (1993) Morphology of conductive, solution-processed blends of polyaniline and poly(methyl methacrylate). *Synthetic Metals*, **53**(3), 293-301.
7. Stenger-Smith, J. D. (1998) Intrinsically electrically conducting polymers: Synthesis, characterization & their application. *Progress in Polymer Science*, **23**, 57-79.
8. Shirakawa, H., Louis, E. J., Mac Diarmid, A. G., Chiang, C. K., and Heeger, A. J. (1977) Synthesis of electrically conducting organic polymers: Halogen derivatives of polyacetylene, $(CH)_x$. *Journal of the Chemical Society, Chemical Communications*, **47**(4), 578-580.
9. Salanck, W. R., Clark, D. T., and Smaullsen E. J. (1991) Science and Application of Conducting Polymers, Adam Hilger, UK.
10. Kinlen, P. J., Silverman, D. C., and Jeffreys, C. R. (1997) Corrosion protection using polyanujne coating formulations. *Synthetic Metals*, **85**, 1327-1332.
11. Nezakati, T., Seifalian, A., Tan, A., and Seifalian, A. M. (2018) Conductive polymers: Opportunities and challenges in biomedical applications. *Chemical Reviews*, **118**(14), 6766–6843.
12. da Silva A. C., and Córdoba de Torresi, S. I., (2019) Advances in conducting, biodegradable and biocompatible copolymers for biomedical applications. *Frontiers in Materials*, **6**, 98.
13. Stejskal, J., Hlavatá, D., Holler, P., Trchová, M., Prokeš, J., and Sapurina, I. (2004) Polyaniline prepared in the presence of various acids: a conductivity study. *Polymer International*, **53**, 294-300.

14. Cao, Y., Qiu, J., and Smith, P. (1995) Effect of solvents and co-solvents on the processibility of polyaniline: I. solubility and conductivity studies. *Synthetic Metals*, **69**, 187–190.
15. Street, G. B., and Clarke, T. C. (1981) Conducting polymers: A review of recent work. *IBM Journal of Research and Development*, **25**, 51–57.

Chapter 7

Magnetism in Materials

Introduction

According to **National Geographical Society Encyclopedia** defini-tion, Magnetism is the force exerted by magnets when they attract or repel each other. Magnetism is caused by the motion of electric charges. Every substance is made up of tiny units called atoms. Each atom has electrons, particles that carry electric charges. Their move-ment generates an electric current and causes each electron to act like a microscopic magnet. In most substances, equal numbers of elec-trons spin in opposite directions, which cancels out their magnetism. In substances such as iron, cobalt, and nickel, most of the electrons spin in the same direction. This makes the atoms in these substances strongly magnetic—but they are not yet magnets. To become mag-netized, another strongly magnetic substance must enter the mag-netic field of an existing magnet. The magnetic field is the area around a magnet that has magnetic force. All magnets have north and south poles. Opposite poles are attracted to each other, while the same poles repel each other. When you rub a piece of iron along a magnet, the north-seeking poles of the atoms in the iron line up in the same direction. The force generated by the aligned atoms creates a mag-netic field. The piece of iron has become a magnet.

7.1 Terminology Involved in Magnetism

There are numerous terms associated with magnetism and before understanding various types of magnetic materials, it becomes man-datory to know them. If we take **magnetic field** as H and **magnetic moment** per unit volume as I, then **magnetic induction**, B for a sub-stance placed in this magnetic field is given by,

$$B = H + 4\pi I$$

where magnetic field, H is defined as a region around a magnetic ma-terial or a moving electric charge within which the force of mag-netism acts. **Magnetic induction**, B is the process by which an object or material is magnetized by an external magnetic field and magnetic moment, I is the measure of the object's tendency to align with a magnetic field. **Magnetic permeability**, relative increase or decrease in the resultant magnetic field inside a material compared with the magnetizing field in which the given material is located; or the property of a material that is equal to the magnetic flux density B established within the material by a magnetizing field divided by the magnetic field strength H of the magnetizing field. **Magnetic permeability μ is** given as

μ = B/H

Magnetic susceptibility is a measure of how much a material will become magnetized in an applied magnetic field. It is the ratio of magnetic moment per unit volume, I to the applied magnetic field, H.

Magnetic susceptibility = I/H

7.2 Classification of Magnetism

There are various types of magnetic materials viz., **ferromagnetic** Materials are those materials which are highly attracted to magnets and can be permanently magnetized. The relative permeability is much greater than unity and they have a high susceptibility. For example, Cr, Ni, Fe, Mn and Co are ferromagnetic.

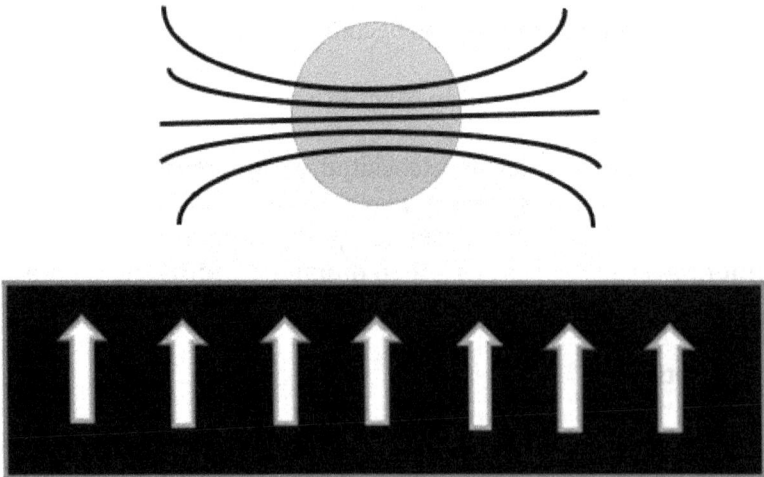

Figure 7.1 Ferromagnetic material.

54

Paramagnetic materials are those materials which are very weakly attracted by the poles of a magnet and fail to retain permanent magnetism. The relative permeability is slightly greater than unity and they attract the lines of forces weakly. For example, Ca, Pt, Al, wood, O_2 are paramagnetic.

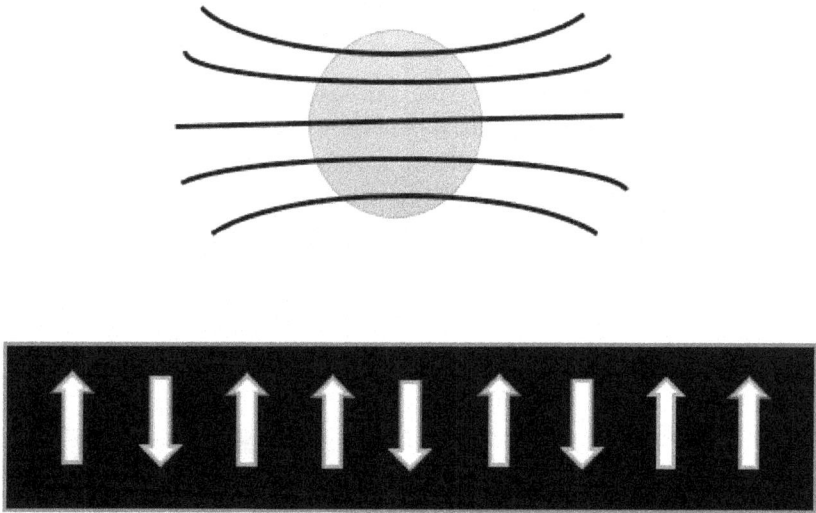

Figure 7.2 Paramagnetic material.

Diamagnetic Materials are repelled by a magnetic field; an applied magnetic field creates an induced magnetic field in them in the opposite direction, causing a repulsive force. Diamagnetic materials have a relative magnetic permeability that is less than or equal to 1, and therefore a magnetic susceptibility less than or equal to 0 Diamagnetic materials include water, wood, most organic compounds such as petroleum and some plastics, and many metals including copper, particularly the heavy ones with many core electrons, such as mercury, gold and bismuth. Diamagnetic and paramagnetic materials are considered nonmagnetic because the magnetizations are relatively small and persist only while an applied field is present. If magnetic susceptibility is negative, the material is diamagnetic. In this case, the magnetic field in the material is weakened by the induced magnetization. Diamagnetic materials are repelled by magnetic fields.

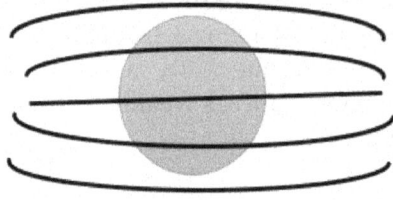

Figure 7.3 A diamagnetic material.

Generally, nonmagnetic materials are said to be para- or diamagnetic because they do not possess permanent magnetization without external magnetic field. **Antiferromagnetic materials** have a tendency for the intrinsic magnetic moments of neighboring valence electrons to point in opposite directions. When all atoms are arranged in a substance so that each neighbor is anti-parallel, the substance is antiferromagnetic. Antiferromagnets have a zero net magnetic moment, meaning that no field is produced by them. Manganese oxide (MnO) is one material that displays this behavior.

(a)

(b)

Figure 7.4 Ferrimagnetic arrangement (a) and antiferromagnetic arrangement (b).

A **ferrimagnetic** material is one that has populations of atoms with opposing magnetic moments, as in anti-ferromagnetism; however, in ferrimagnetic materials, the opposing moments are unequal, and a spontaneous magnetization remains. Ferromagnetic, ferrimagnetic, or antiferromagnetic materials possess permanent magnetization even without external magnetic field and do not have a well-defined zero-field susceptibility. Ferrites (widely used in household products such as refrigerator magnets) are usually ferrimagnetic ceramic compounds derived from iron oxides. Magnetite (Fe_3O_4) is a famous example.

In addition to susceptibility differences, the different types of magnetism can be distinguished by the structure of the magnetic dipoles in regions called **domains**. A domain wall is an interface separating magnetic domains. It is a transition between different magnetic moments and usually undergoes an angular displacement of 90° or 180°. A **domain wall** is a gradual reorientation of individual moments across a finite distance. The energy of a domain wall is simply the difference between the magnetic moments before and after the domain wall was created. This value is usually expressed as energy per unit wall area. Each domain consists of magnetic moments that are aligned, giving rise to a permanent net magnetic moment per domain. There are two types of domain walls viz., **Bloch wall (A)** - A Bloch wall is a narrow transition region at the boundary between magnetic domains, over which the magnetization changes from its value in one domain to that in the next. In a Bloch domain wall, the magnetization rotates about the normal of the domain. Bloch domain walls appear in bulk materials, i.e., when sizes of magnetic material are considerably larger than domain wall width.

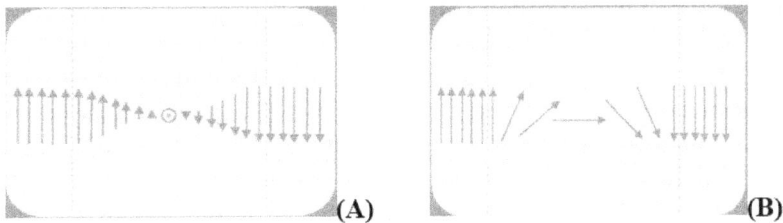

Figure 7.5 Domain wall – Bloch (A) and Neel (B) wall.

Neel wall (B) - A Neel wall is a narrow transition region between magnetic domains. In the Neel wall, the magnetization smoothly rotates from the direction of magnetization within the first domain to the direction of magnetization within the second. Neel walls are the common magnetic domain wall type in very thin films where the exchange length is very large compared to the thickness.

Domains exist even in absence of external field. In a material that has never been exposed to a magnetic field, the individual domains have a random orientation. This type of arrangement represents the lowest free energy. When the bulk material is un-magnetized, the net magnetization of these domains is zero, because adjacent domains may be orientated randomly in any number of directions, effectively canceling each other out. The average magnetic induction of a ferromagnetic material is intimately related to the domain structure. When a magnetic field is imposed on the material, domains that are nearly lined up with the field grow at the expense of unaligned domains. This process continues until only the most favorably oriented domains remain. In order for the domains to grow, the Bloch walls must move, the external field provides the force required for this moment. When the domain growth is completed, a further increase in the magnetic field causes the domains to rotate and align parallel to the applied field. At this instant material reaches saturation magnetization and no further increase will take place on increasing the strength of the external field. Under these conditions the permeability of these materials becomes quite small.

7.3 Factors Affecting Magnetism

Temperature does have a definite effect on a materials' magnetic behavior. With rising temperature, magnitude of the atom thermal vibrations increases. This may lead to more randomization of atomic magnetic moments as they are free to rotate. Usually, atomic thermal vibrations counteract forces between the adjacent atomic dipole moments, resulting in dipole misalignment up to some extent both in presence and absence of external field. As a consequence of it, saturation magnetization initially decreases gradually, then suddenly drops to zero at a temperature called **Curie temperature**, Tc. The magnitude of the Curie temperature is dependent on the material. For example: for cobalt – 1120 °C, for nickel – 335 °C, for iron – 768 °C, and for Fe_3O_4 – 585 °C.

7.4 Magnetism in Various Materials

There are various materials that exhibit magnetism. Some of them are magnetic at all times. Other, like stainless steel, have magnetic properties only with a certain chemical composition. **Iron** is an extremely well-known ferromagnetic metal. It is, in fact, the strongest ferromagnetic metal. It forms an integral part of the earth's core and imparts its magnetic properties to our planet. That is why the Earth acts as a permanent magnet on its own. There are many aspects that contribute to iron's magnetism. In addition to its net electron spin at the atomic level, its crystalline structure also plays an important role. Without it, iron would not be a magnetic metal. Iron is ferromagnetic in its body-centred cubic (bcc) alpha-Fe structure (stable below 910°C). At the same time, it does not show magnetism in face-centred cubic (fcc) gamma-Fe structure (910°C and 1400°C). Beta-Fe structure (between 768° and 910° C), for example, displays paramagnetic tendencies.

Nickel is another popular magnetic metal with ferromagnetic properties. Like iron, its compounds are present in the earth's core. Historically, nickel has been used to make coins. Today, nickel finds use in batteries, coatings, kitchen tools, phones, buildings, transport and jewelry. A large portion of nickel is used to manufacture ferronickel for stainless steel. Because of its magnetic properties, nickel is also part of Alnico magnets (made of aluminium, nickel, and cobalt). These magnets are stronger than rare-earth metal magnets but weaker than iron-based magnets.

Cobalt is an important ferromagnetic metal. Cobalt can be used to produce soft as well as hard magnets. Soft magnets that use cobalt have advantages over other soft magnets. Thus, they can be used for high-temperature applications (up to under 500° Celsius). Cobalt with its alloys is used in hard disks, wind turbines, MRI machines, motors, actuators, and sensors.

Steel also displays ferromagnetic properties as it is derived from iron. Most steels will be attracted to a magnet. If needed, steel can also be used to make permanent magnets. Some stainless steels are magnetic, and some are not. An alloy steel becomes a stainless steel if it has at least 10.5% of chromium in it. Due to the varying chemical compositions, there are different types of stainless steel. Ferritic and martensitic stainless steels are magnetic due to their iron composition and molecular structure. Austenitic steels, on the other hand, do not display ferromagnetic properties because of a different molecular structure. This makes the suitable for use in MRI machinery. The structural difference derives from the amount of nickel. It strengthens the oxide layer for better protection against corrosion but also changes the structure of stainless steel.

Along with the above-mentioned metals, compounds of some **rare earth elements** also have excellent ferromagnetic properties. Gadolinium, samarium and neodymium are all examples of magnetic rare earth metals. Various magnets with different properties can be manufactured using the above metals in combination with iron, nickel and cobalt. These magnets come with specific properties necessary for certain applications. For example, samarium-cobalt magnets are present in turbomachinery, high-end electric motors, etc.

The **spinels** can be ferromagnetic or anti ferrimagnetic depending on the structure and nature of metal ions. The unpaired spins of metal ions are coupled through shared oxide ions by super exchange process. In ferrite spinels, the spins of electrons at tetrahedral sites have one orientation, whereas the spins of electrons at octahedral sites have opposite orientation (as explained in Chapter 4). If the number of spins in these two types of sites is equal, then that spinel will be antiferromagnetic. Otherwise, it will be ferromagnetic. E.g., The inverse spinels - Fe_3O_4, $NiFe_2O_4$ and $CoFe_2O_4$ are ferrimagnetic. In these cases, the spins of trivalent ions are cancelled out since half of them belong to tetrahedral sites and the other half belong to octahedral sites. However, the spins of divalent ions are not cancelled. Hence, they show ferrimagnetism. Antiferromagnetic substances show weak magnetism, whereas diamagnetic substances show no magnetism at all.

7.5 Uses of Magnetism in Materials

In Transformers and Motor Cores

For materials which have large power handling capacity and incur low losses can be used as transformer and motor cores. They should also have high permeability, should also be able to magnetize easily in low magnetic fields and in which domain walls are able to migrate easily. Hence, keeping in mind these characteristics, ferro- and ferrimagnetic materials are most appropriate to be used as transformer and motor cores.

In Magnetic Storage Devices

Magnetic storage or magnetic recording is the storage of data on a magnetized medium. Magnetic storage uses different patterns of magnetization in a magnetizable material to store data and is a form of non-volatile memory. The information is accessed using one or more read/write heads. Magnetic storage media, primarily hard disks, are widely used to store computer data as well as audio and video signals. In the field of computing, the term magnetic storage is preferred and in the field of audio and video production, the term magnetic recording is more commonly used. The distinction is less technical and more a matter of preference. Other examples of magnetic storage media include floppy disks, magnetic tape, and magnetic stripes on credit cards.

Permanent Magnets

An object, made from a material that can be magnetized and can generate its own persistent magnetic field can serve as a Permanent Magnet. Materials that can be magnetized, which are also the ones that are strongly attracted to a magnet, are called ferromagnetic (or ferrimagnetic). These include the elements iron, nickel and cobalt and their alloys and some alloys of rare-earth metals. Although ferromagnetic (and ferrimagnetic) materials are the only ones attracted to a magnet strongly enough to be commonly considered magnetic, all other substances respond weakly to a magnetic field, by one of several other types of magnetism. An everyday example of a permanent magnet is a refrigerator magnet used to hold notes on a refrigerator door.

References

1. Mohn, P. (2006) *Magnetism in the Solid State: An Introduction*, 2nd volume, Springer, Germany.
2. *Magnetic Oxides*, Part 1 and 2, Craik, D. J. (ed.), Wiley, USA (!975).
3. Earnshaw, A. (1968), *Introduction to Magnetochemistry*, Academic Press, USA.
4. Leslie-Pelecky, D. L., and Rieke, R. D. (1996) Magnetic properties of nanostructured materials. *Chemistry of Materials*, **8**(8), 1770-1783. DOI: 10.1021/cm960077f.

5. Zijlstra, H. (1967) *Experimental Methods in Magnetism*, North-Holland Publication Company, Holland.
6. Shull, C. G., Strauser, W. A., and Wollan, E.O. (1951) Neutron diffraction by paramagnetic and antiferromagnetic substances. *Physical Review*, **83**, 333.
7. Standley, K. J. (1972) *Oxide Magnetic Materials*, Clarendon Press, UK.
8. Taylor, K. N. R. (1970) The rare earth metals. *Contemporary Physics*, **11**, 423.
9. Tebble, R. S., and Graik, D. J. (1979) *Magnetic Materials*, Wiley, USA.
10. *Treatise on Solid State Chemistry*, Hannay, N. B. (ed.). volume 1, Plenum Press, USA (1973).

Chapter 8

Organic Charge Transfer Complexes

Introduction

A charge-transfer (CT) complex is more of an **electron-donor-acceptor complex** which may be elaborated as an association of two or more molecules, in which a fraction of electronic charge is transferred between the molecular entities. The molecular complex is stabilized by the resulting electrostatic attraction. The **electron donor** (ED) is the source molecule from which the charge is transferred and **electron acceptor** (EA) is the receiving species. A number of organic compounds form electron-donor-acceptor complexes (EDA complexes). Characteristically, these charge transfer complexes crystallize in stacks of alternating donor and acceptor molecules, i.e. A-B-A-B. Such compounds tend to behave in a highly conducting manner and sometimes as superconducting material at low temperatures.

Figure 8.1 Block diagram of mechanism of charge transfer complex formation.

For the first time it was discovered in 1973 that a combination tetracyanoquinodimethane (TCNQ) and tetrathiafulvalene (TTF) forms a strong charge-transfer complex, henceforth referred to as TTF-TCNQ.

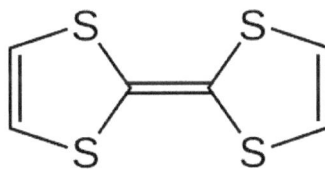

(TTF)

63

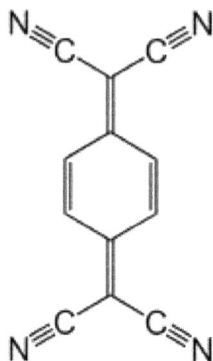

(TCNQ)

The solid shows almost metallic electrical conductance and was the first discovered purely organic conductor. In a TTF-TCNQ crystal, TTF and TCNQ molecules are arranged independently in separate parallel-aligned stacks, and an electron transfer occurs from donor (TTF) to acceptor (TCNQ) stacks. Hence, electrons and electron holes are separated and concentrated in the stacks and can traverse in a one-dimensional direction along the TCNQ and TTF columns, respectively, when an electric potential is applied to the ends of a crystal in the stack direction. Some other examples of electron donors and electron acceptors are shown below.

8.1 Electron Donor Molecules

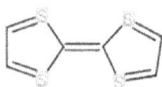

Tetrathiafulvalene
(TTF)

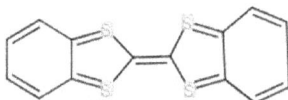

Dibenzotetrathiafvalene
(DBTTF)

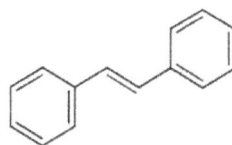

Trans-Stilbene

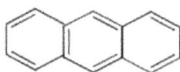

Napthalene

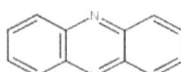

Anthracene

Phenazine

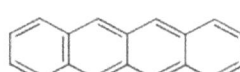

Tetraccene

64

8.2 Electron Acceptor Molecules

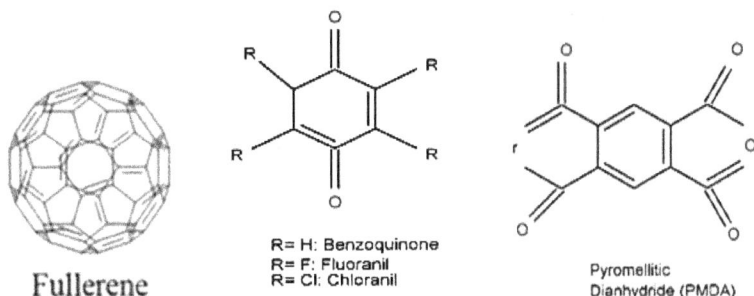

Fullerene

R= H: Benzoquinone
R= F: Fluoranil
R= Cl: Chloranil

Pyromellitic
Dianhydride (PMDA)

8.3 Mechanism of Reaction Between p-Nitroaniline and Chloranilic Acid

In one such example, the procedure of formation of a charge transfer complex is depicted by the interaction between p-nitroaniline and chloranilic acid. In this case, Chloranilic acid acts as an electron donor while p-nitroaniline acts as an electron acceptor.

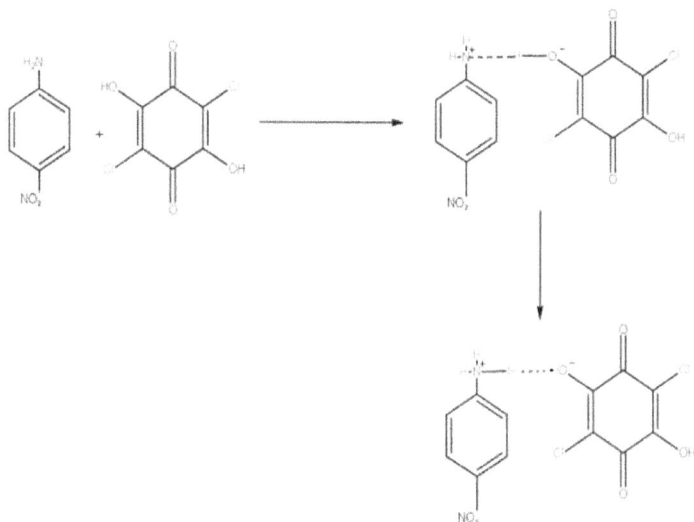

Figure 8.2 Procedure of formation of a charge transfer complex from p-nitroaniline and chloranilic acid.

8.4 Mechanism of Reaction Between DAP and CLA

New charge transfer complex between the electron donor 2,3-diaminopyridine (DAP) with the electron acceptor chloranilic acid (CLA) has been synthesized which is in salt crystals and generates positive and negative radicals in the presence of polar solvents to ease the conductivity. The analysis of DFT results strongly confirmed the high stability of the formed complex based on existing charge transfer beside proton transfer hydrogen bonding concordant with experimental results.

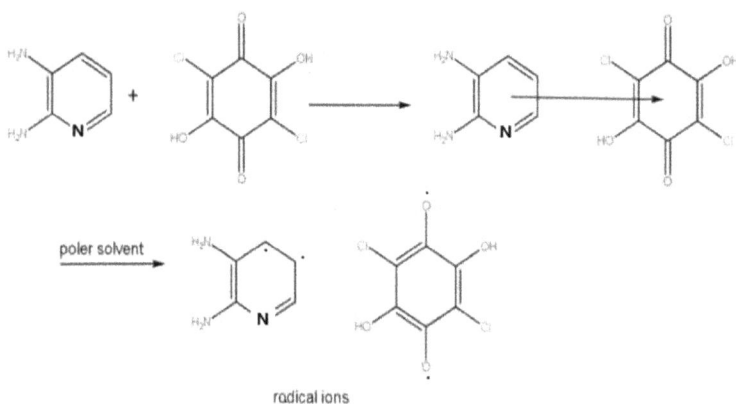

Figure 8.3 Procedure of formation of a charge transfer complex from 2,3-diaminopyridine and chloranilic acid.

8.5 Mechanism of Reaction Between p-Toluidine and Chloranilic Acid

Another example also shows the formation of a charge transfer complex formed by the interaction of p-toluidine (π electron donor) and chloranilic acid (π electron acceptor). These complexes arise from the interaction of a Lewis acid-Lewis base where bond between the components of complex arise due to partial transfer of an electron from a base to the empty orbital of an acid. Charge transfer complex have a transition between an excited molecular state and a ground state.

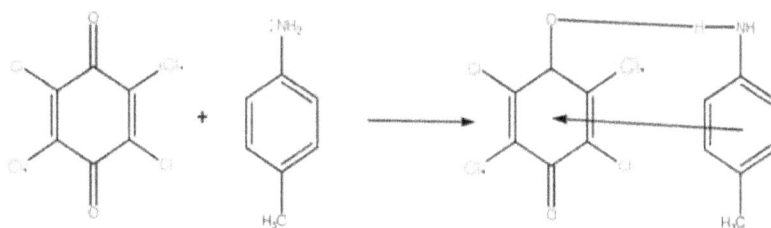

Figure 8.4 Procedure of formation of a charge transfer complex from p - toluidine and chloranilic acid.

8.6 Mechanism of Reaction Between 2,3-Dichloro-5,6-dicyanobenzoquinone and Substituted Phenol

2, 3-dichloro-5, 6-dicyanobenzoquinone accepts electrons from sub-stituted phenol which acts as electron donor. They transfer electrons to form charge transfer complex and, hence, remain in association.

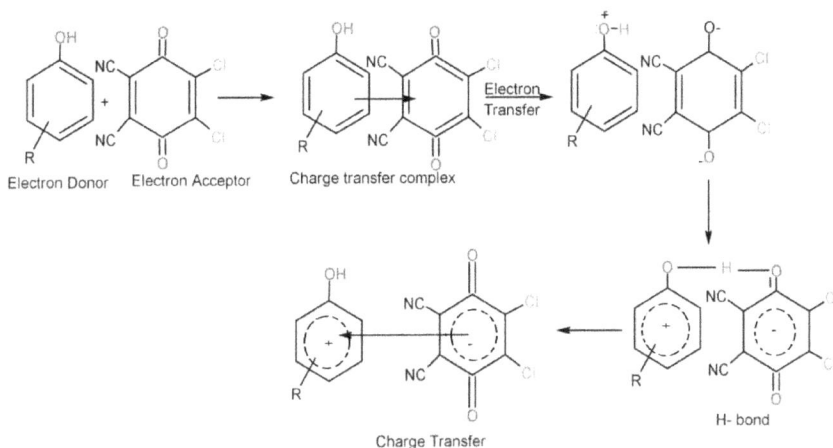

Figure 8.5 Procedure of formation of a charge transfer complex from 2, 3-dichloro-5, 6-dicyanobenzoquinone and substituted phenol.

8.7 Applications of Charge Transfer Complexes

Charge transfer complexes are widely used as organic conductors and superconductors. An example of Superconductivity is exhibited by   di(2,3,6,7-tetramethyl-1,4,5,8-tetraselenafulvalenium) hexafluorophosphate, (TMTSF)$_2$PF$_6$, which is a semi-conductor at ambient conditions, shows superconductivity at low temperature (critical temperature) and high pressure: 0.9 K and 12 kbar. Current advances in molecular engineering have led to the development of switches and memory devices, transistors, photovoltaic cells, etc.

References

1. Anderson, P. W., Lee, P. A. and Saitoh, M. (1973) Remarks on giant conductivity in TTF-TCNQ. *Solid State Communications*, **13**(5), 595–598.
2. Kochi, J. K. (1988) Electron transfer and charge transfer: Twin themes in unifying the mechanisms of organic and organometallic reactions. *Angewandte Chemie International Edition*, **27**(10), 1227–1266.
3. Khan, S. A. (2014) Charge-transfer complexes: A short review. *International Journal of Emerging Technologies and Innovative Research*, **1**(2), 52-58.
4. Ishaat, M., Khan, M., Ahmad, A., Miyan, L., Ahmad, M., and Azizc, N. (2017) Synthesis of charge transfer complex of chloranilic acid as acceptor with p-nitroaniline as donor: Crystallographic, UV–visible spectrophotometric and antimicrobial studies. *Journal of Molecular Structure*, **1141**, 687-697.
5. Taku, M. (1965) A study of the charge-transfer complexes. II. The complexes of pyromellitic dianhydride with polycyclic aromatic compounds. *Bulletin of the Chemical Society of Japan*, **38**(12), 2110-2114.
6. Al-Ahmary, K. M., Habeeb, M. M., and Al-Obidan, A. H. (2018) Charge transfer complex between 2,3-diaminopyridine with chloranilic acid. Synthesis, characterization and DFT, TD-DFT computational studies. *Spectrochimica Acta Part A: Molecular and Biomolecular Spectroscopy*, **196**, 247-255.
7. Sharma, R., Paliwal, M., Singh, S., Ameta, R. and Ameta, S. C. (2008) Synthesis, characterization and electrical conductivity of charge transfer complex of p-toluidine and chloranil. *Indian Journal of Chemical Technology*, **15**, 613-616.
8. Yamaguchi, S., Viands, C. A., and Potember, R. S. (2001) Atomic force microscopy observation of the morphology of tetracyanoquinodimethane (TCNQ) deposited from solution onto the atomically smooth native oxide surface of Al(111) films. *Journal of Vacuum Science and Technology*, **B9**, 1129.
9. Goetz, K. P., Vermeulen, D., Payne, M. E., Kloc, C., McNeil, L. E., and Jurchescu, O. D. (2014) Charge-transfer complexes: New

perspectives on an old class of compounds. *Journal of Materials Chemistry C*, **2**(17), 3065–3076.

10. Lee, C. S. (2014) Application of charge transfer complexes in organic optoelectronic devices. *Solid-State and Organic Lighting*, **2**(4), 45-51.

11. Zhang, J., Xu, W., Sheng, P., Zhao, G., and Zhu, D. (2017) Organic donor-acceptor complexes as novel organic semiconductors. *Accounts of Chemical Research*, **50**, 1654– 1662.

Chapter 9

New Superconductors

Introduction

Superconductors are those materials that offer little or no resistance to electricity when we cool them down to very low temperatures. We know that there are two types of materials: **Conductors** (such as metals) carry electricity well, while **insulators** (such as plastics) barely let it pass through them at all. But it will be more accurate to say that all materials conduct electricity, under the right conditions, but some conduct more easily than others. When a metal conducts electricity well, it means that the metal offers little or no resistance when you try to make a current flow through it; when plastics insulate well, it means that they put up high resistance to electric currents. Resistance is the keyword while dividing materials into "conductors" and "insulators".

Resistance changes with change in the temperature. For example, Gold is one of the best conductors which means it shows very little resistance to electricity, but on increasing its temperature, it puts up much more resistance. The reason for this is, the higher the temperature, the more thermal vibrations occur inside the gold's crystalline structure and the harder electrons (the negatively charged particles inside atoms that carry electric currents) will find it to flow through. Alternatively, when we cool gold down, we reduce the vibrations and make it easier for electrons to flow. At low temperatures, impurities and defects in the material cause most of the resistance. A fairly complex mathematical equation called Matthiessen's rule helps to figure out the total resistance of a material at any given temperature by summing the various effects.

9.1 Working of a Superconductor

Dutch physicist, **Heike Kamerlingh Onnes** in 1911 was the first one to probe into the temperature dependence of conductivity. During his experiments, when he cooled a wire made of mercury to a very low temperature of 4.2K, he observed that its electrical resistance suddenly disappeared i.e., he discovered superconductivity. However, it

71

was also coupled with the fact that if strong magnetic field is applied to mercury, the superconductivity vanished immediately.

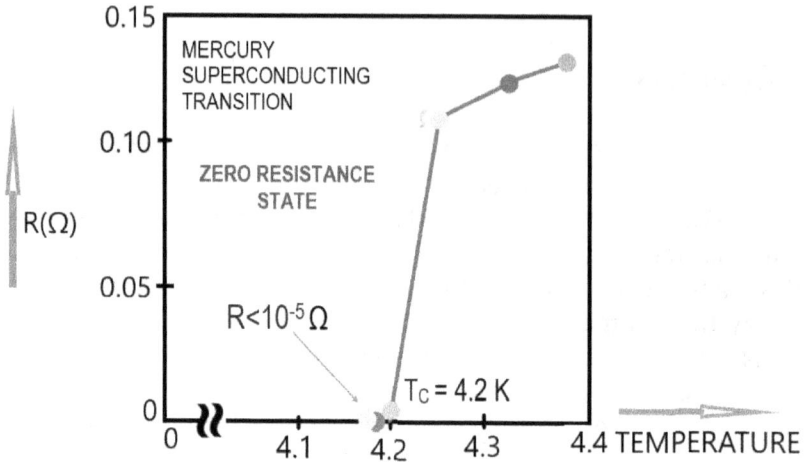

Figure 9.1 Superconductivity in mercury.

German physicists, **Karl Meissner and Robert Ochsenfeld** extended the concept of superconductivity that was proposed by Onnes. A superconductor is diamagnetic: it refuses to let magnetism to penetrate inside it. Stand a superconductor in a magnetic field and you'll make electric currents flow through its surface. These currents create a magnetic field that exactly cancels the original field trying to get inside the superconductor and repelling the magnetic field outside. This is known as the **Meissner effect** and it explains how a superconductor floats in a magnetic field. They have zero resistivity and perfect diamagnetism.

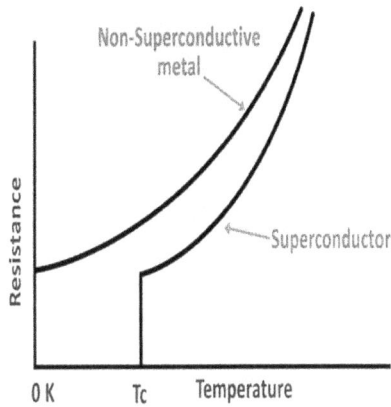

Figure 9.2 Graphical representation of superconductors.

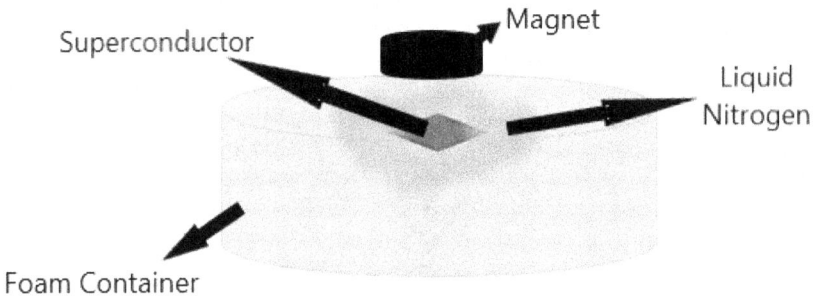

Figure 9.3 Working of a superconductor.

A superconductor works on the **BCS (Bardeen-Cooper-Schrieffer) theory,** the name given in honor of its three discovers, John Bardeen, Leon Cooper and Robert Schrieffer in 1972. It explains that materials suddenly become "superb conductors" when the electrons inside them join forces to make what are called **Cooper pairs** (or BCS pairs). Normally, the electrons that carry electricity through a material are scattered about by impurities, defects, and vibrations of the material's crystal lattice. And this is known as electrical resistance. But at low temperatures, when the electrons join together in pairs, they can move more freely without being scattered in the same way.

Getting into the detail of the theory, it is assumed that there is some attraction between electrons, which can overcome the Coulomb repulsion and the nearby positive charges in the lattice are attracted by these electrons moving through a conductor. This causes deformation of the lattice with respect to another electron, having opposite spin, which moves into the region of higher positive charge density. These two electrons then become correlated and behave as a pair (Cooper pair).

The lattice vibration of type I superconductor is minimized on being cooled below critical temperature. When an electron passes by this structure, it attracts the positive lattice and thereby brings about distortion in the structure, due to which a phonon is released, and a net positive area is created. A phonon is a packet of vibrational energy. Another electron is attracted to this area, absorbing the phonon providing it sufficient energy to overcome electrostatic repulsion and joins with the initial electron, forming a Cooper pair. The Cooper pair acts as one particle and can move through the lattice unimpeded, thus there is zero electrical resistance.

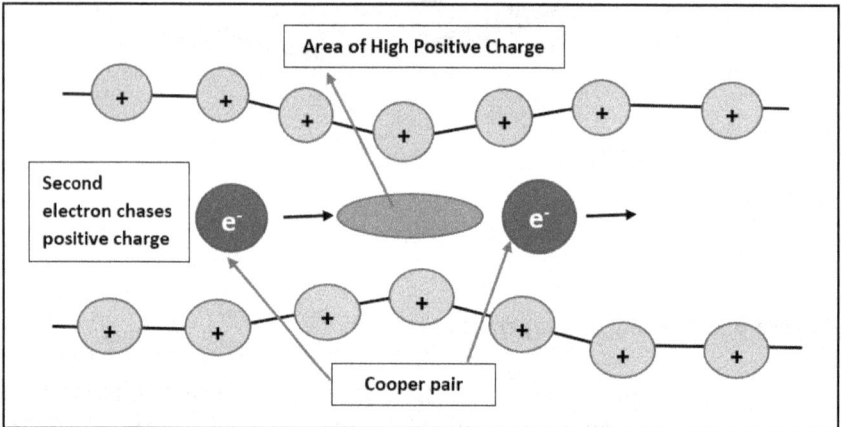

Figure 9.4 Cooper pair.

As there are a lot of such electron pairs in a superconductor, these pairs overlap very strongly and form a highly collective condensate. Since this condensate works as a system hence the breaking of one pair will change the energy of the entire condensate which implies that the energy required to break any single pair is related to the energy required to break all of the pairs. Due to pairing, the energy barrier is increased, it cannot be crossed by the oscillating atoms in the conductor, which are small at low temperatures and are not enough to affect the condensate. Thus, the electrons stay paired together and resist all hindrances, and the electron flow, as a whole does not experience resistance. Thus, the collective behavior of the condensate is the key for superconductivity.

9.2 Superconducting Materials

Not all materials show superconductivity. Apart from mercury, the original superconductor, there are about 25 other elements showing this property which are mostly metals, semimetals, or semiconductors, though it's also been discovered in thousands of compounds and alloys. Each different material becomes a superconductor at a slightly different temperature, known as its **critical temperature or Tc**. These superconduct only within a few degrees of **absolute zero** (the lowest theoretically possible temperature: −273.15°C, −459.67°F, or 0K), which means the benefit gained from their lack of resistance is lost from the energy provided in having to cool them down. This is the main reason why superconductors have yet to make a really big impact on the world, despite being discovered almost a century ago.

9.3 Types of Superconductors

A superconductor with little or no magnetic field within it is said to be in the Meissner state (as described earlier). The breakdown of Meissner state occurs when the applied magnetic field is too large. On the basis of how this breakdown occurs, Superconductors can be divided into two classes.

1. **Type I superconductors**: They are the materials in which the transition from a superconducting state to a normal state due to the external magnetic field is sharp and abrupt i.e., Superconductivity is **instantly** destroyed with the rise in the strength of the applied magnetic field.

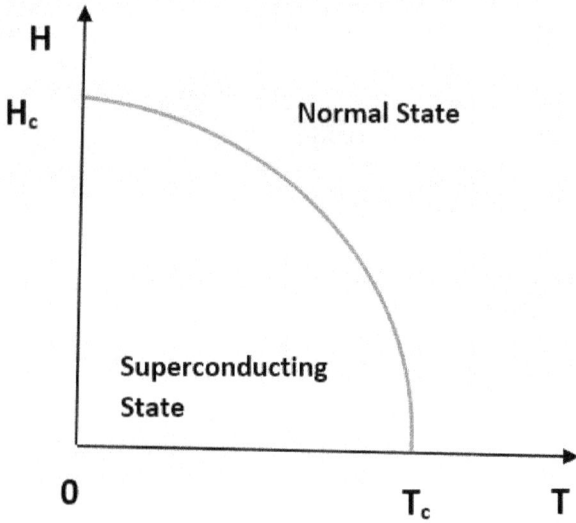

Figure 9.5 Critical magnetic field vs temperature (in Kelvin) for Type I superconductors.

They have a **low critical temperature** (typically in the range of 0K to 10K) and a **low critical magnetic field (**hence cannot be used for preparing electromagnets which produce strong magnetic field.**)**. Type I Superconductors perfectly **obey the Meissner effect** i.e., magnetic field cannot penetrate inside the material. Since Type I Superconductors easily lose the superconducting state by low-intensity magnetic field hence they are also known as **soft superconductors**. Type 1 Superconductors are completely **diamagnetic** and their superconductivity can be explained by BCS Theory This type of superconductivity is normally exhibited by pure metals, **e.g., aluminium, lead, and mercury.** The only alloy known which exhibits type I superconductivity is $TaSi_2$. The covalent superconductor **SiC-B** (silicon carbide heavily doped with boron) is also type-I superconductor.

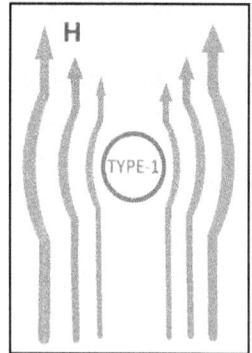

2. **Type II superconductors**: They are the materials in which the transition from a superconducting state to a normal state due to the external magnetic field is **gradual**. At lower critical magnetic field (H_1), type-II superconductor starts losing its superconductivity. At upper critical magnetic field (H_2), type-II superconductor completely loses its superconductivity. The state between lower critical magnetic field and upper magnetic field is known as an intermediate state or **mixed state**.

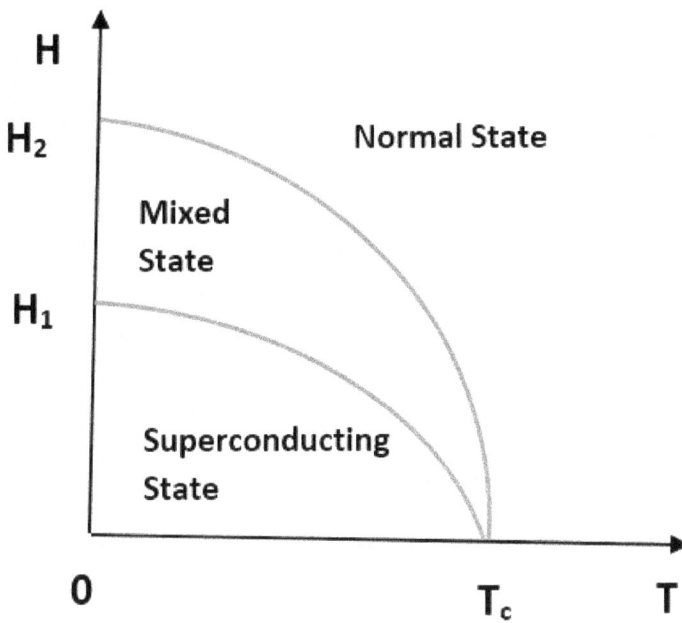

Figure 9.6 Critical magnetic field vs temperature (in Kelvin) for Type II superconductors.

They have a **high critical temperature** (mostly greater than 10K) and a **high Critical magnetic field** (hence can be used for preparing electromagnets which produce strong magnetic field.). Type II Superconductors **partly obey the Meissner effect** i.e., Magnetic field can penetrate inside the material. Since Type II Superconductors does not easily lose the superconducting state by external magnetic field, hence they are also known as **hard superconductors**. Type 1I Superconductors are not completely **diamagnetic** and their superconductivity cannot be explained by BCS Theory This type of superconductivity is normally exhibited by pure metals, **e.g., alloys and complex oxides of ceramics, NbTi, Nb$_3$Sn.**

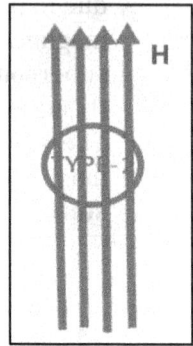

Details of Type II or High-Temperature Semiconductors

For many years, scientists assumed superconductivity could happen only at very low temperatures. Then, in 1986, two European scientists working for IBM, German physicist J. Georg Bednorz and Swiss physicist K. Alex Müller, discovered a ceramic cuprate (a material containing copper and oxygen) that could become a superconductor at much higher temperatures (−238°C, −396°F, or 35K). Other scientists have since found materials that show superconductivity at even higher temperatures and the record is currently held by a material called mercury thallium barium calcium copper oxide (Hg$_{12}$T$_{13}$Ba$_{30}$Ca$_{30}$Cu$_{45}$O$_{125}$), which superconducts at −135°C (−211°F or 138K) and was

The discovery of so-called high-temperature superconductors (HTS) moved research on enormously. The original superconductors needed temperatures within a whisker of absolute zero—and it can reach those only by cooling materials using an expensive coolant gas such as liquid helium. But the high-temperature superconductors can be cooled using liquid nitrogen instead, which is about 10 times cheaper to produce. A lot of applications that weren't economic suddenly became a whole lot more practical when high-temperature superconductors were discovered.

In 2020, scientists made another breakthrough with the discovery of materials that superconduct at everyday temperatures (around 15°C or 59°F)—but there's still an awkward catch: this happens only at enormously high pressures (over 2 million times atmospheric pressure). The challenge now is to find materials that superconduct at everyday temperatures and pressures.

9.4 Applications of Superconductors

Applications in Areas of Magnetism

The most widespread practical use for superconductors at the moment is in body scanners, based on NMR (nuclear magnetic resonance). When we direct an intense magnetic field at an atom, we can make its nucleus resonate and give off radio waves. In a body scanner, superconducting magnets make the powerful magnetic field, which causes atoms inside the patient's body to give off radio waves. As the scanner spins around, it picks up these waves and turns them into an image of inside of the patient's body. This is called **magnetic resonance imagery (MRI)** and it currently uses low-temperature superconductors.

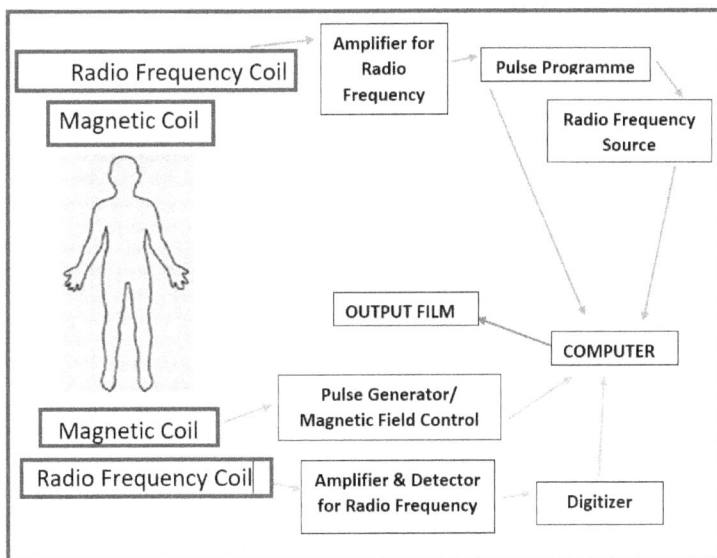

Figure 9.7 Block diagram of the working principle of magnetic resonance imagery (MRI) analysis.

Superconducting magnets are well known for their role in particle accelerators as in CERN's **Large Hadron Collider** (LHC). If charged particles (like bits of atoms) move through a magnetic field, they bend round in a curve. This can be used to accelerate them to extremely high speeds and energies so, when they collide, they hit and move apart and generate new particles that help to understand the detailed structures. The LHC, for example, uses over one thousand magnets made from a niobium-titanium alloy (Nb-Ti) cooled almost to absolute zero (also types of **low-temperature superconductors**). The magnetic field produced by Collider is over one lakh times greater than Earth's magnetic field.

Floating trains that use **magnetic levitation ("Maglev")** is another application of superconductivity. One exception is the Japanese SCMaglev system, which uses superconducting magnets to float trains and speed them along at up to 603 kph (375 mph). High speed is only one major benefit of maglev trains. As the trains rarely touchthe track, there's very less noise and vibration than conventional trains. Less vibration and friction results in few mechanical breakdowns, resulting in maglev trains to be less likely to encounter weather-related delays.

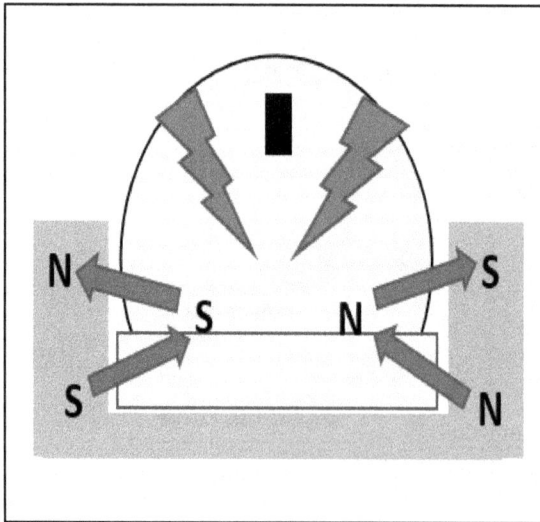

Figure 9.8 Levitation system of floating trains.

Applications in Areas of Electricity

There is no electrical resistance in a superconductor, and therefore no energy loss. There is, however, a maximum supercurrent that can flow, called the critical current. Above this critical current the material is normal. There is one other very important property: when a metal goes into the superconducting state, it expels all magnetic fields, as long as the magnetic fields are not too large. In a **Josephson junction**, the non-superconducting barrier separating the two superconductors must be very thin. Until a critical current is reached, a supercurrent can flow across the barrier; electron pairs can tunnel across the barrier without any resistance. But when the critical current is exceeded, another voltage will develop across the junction. That voltage will depend on time--that is, it is an AC voltage. This in turn causes a lowering of the junction's critical current, causing even more normal current to flow--and a larger AC voltage.

Josephson junctions have been used to make ultrafast logic gates and extremely sensitive magnetism detectors called **SQUID**s (Superconducting Quantum Interference Devices), which are finding their way into all kinds of things from improved MRI brain scanners to supersensitive submarine detectors. These devices are extremely sensitive and very useful in constructing extremely sensitive magnetometers and voltmeters.

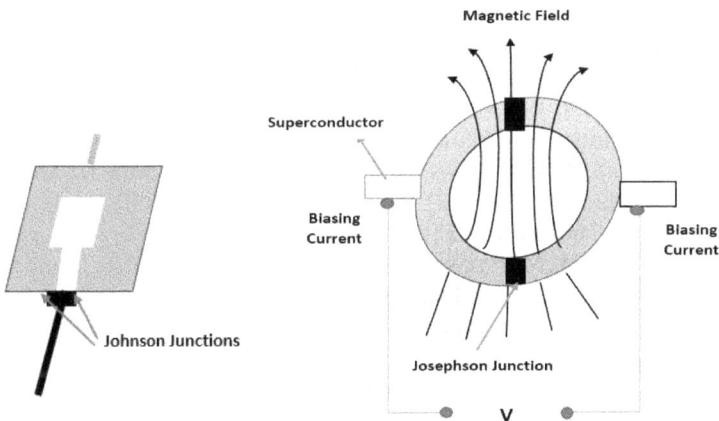

Figure 9.9 Superconducting quantum interference devices.

A **SQUID** consists of a loop with two Josephson junctions interrupting the loop. A SQUID is extremely sensitive to the total amount of magnetic field that penetrates the area of the loop--the voltage that you measure across the device is very strongly correlated to the total magnetic field around the loop.

SQUIDs are being used for research in a variety of areas. Since the brain operates electrically, one can, by sensing the magnetic fields created by neurological currents, monitor the activity of the brain--or the heart. There is a novel use as a SQUID magnetometer for geological research, detecting remnants of past geophysical changes of the earth's field in rocks. Similarly, changes in the ambient magnetic field are created by submarines passing below the surface of the ocean, and hence SQUIDs can be helpful for submarine detection.

References

1. Ogmen, S. (2018) Recent Progress in Superconductor Theory, Materials and Devices, doi: DOI:10.13140/RG.2.2.22452.96646.
2. Romanovskii, V. R., and Rudnev, I. A. (2021) Applied superconductivity and magnetism. *Multidisciplinary Science Journal*, **4**, 82-83.
3. Cohen, M. L. (1988) The new high temperature superconductors. *Transactions on Nuclear Science*, **35**(1), 22-26.
4. Yuan, J., Stanev, V., Gao, C., Takeuchi, I., and Jin, K. (2019) Recent advances in high-throughput superconductivity research. *Superconductor Science and Technology*, **32**(12), 123001.
5. Bardeen, J., Cooper, L. N., and Schrieffer, J. R. (1957) Microscopic theory of superconductivity. *Physical Review*, **106**(1), 162–164.
6. Bardeen, J., Cooper, L. N., and Schrieffer, J. R. (1957) Theory of superconductivity. *Physical Review*, **108**(5), 1175–1204.
7. Gottlieb, U., Lasjaunias, J. C., Tholence, J. L., Laborde, O., Thomas, O. and Madar, R. (1992) Superconductivity in $TaSi_2$ single crystals. *Physical Review B*, **45** (9), 4803–4806.
8. Kriener, M., Muranaka, T., Kato, J., Ren, Z. A., Akimitsu, J., and Maeno, Y. (2008) Superconductivity in heavily boron-doped silicon carbide. *Science and Technology of Advanced Materials*, **9**(4), 044205.
9. Berger, A. (2002) Magnetic resonance imaging. *BMJ*, **5**, 324.
10. Dong, Z., Andrews, T., Xie, C., and Yokoo, T. (2015) Advances in MRI techniques and applications. BioMed Research International, **2015**, Article ID 139043.

11. Wahsner, J., Gale, E. M., Rodríguez-Rodríguez, A., and Caravan, P. (2019) Chemistry of MRI contrast agents: Current challenges and new frontiers. *Chemical Reviews*, **119**(2), 957–1057.
12. Potter, K. (1996) The Large Hadron Collider (LHC) project of CERN. Online: https://cds.cern.ch/record/308243/files/lhc-project-report-36.pdf [accessed 21st August 2021].
13. Rhodes, C. (2013) Large Hadron Collider (LHC). *Science Progress*, **96**, 95-109.
14. Rahee, A., Yu, J., Muhammad, A., and Naveed, J. (2014) Comprehensive study and review on Maglev train system. *Applied Mechanics and Materials,* **615**, 347-351.
15. Pednekar, S., Singh, A., Oza, Y., Awad, R., and Trapathi, P. (2017) Maglev train. *International Journal of Engineering Research and Technology*, **5**(1).
16. Hyung-Woo, L., Ki-Chan, K., and Lee, J. (2006) Review of maglev train technologies. *Transactions on Magnetics*, **42**(7), 1917-1925.
17. Wolf P. (1977) Computer applications of Josephson junctions. In: Superconductor Applications: SQUIDs and Machines, Schwartz, B. B., and Foner, S. (eds.) Springer, USA.
18. Tafuri, F. (2021) Introduction: the Josephson effect and its role in physics. *Journal of Superconductivity and Novel Magnetism*, **34**, 1581–1586.

84

Chapter 10

Fullerenes and Doped-Fullerenes

Introduction

Allotropic forms of carbon include diamond, graphite, amorphous carbon, fullerenes, carbon nanotubes, graphenes and also some more.

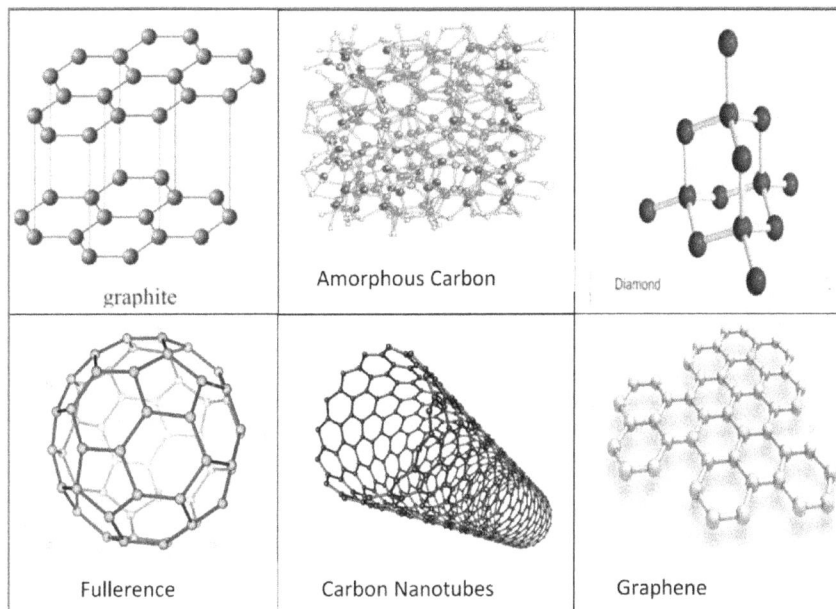

graphite	Amorphous Carbon	Diamond
Fullerence	Carbon Nanotubes	Graphene

10.1 Structure of Fullerene

Getting into the details of fullerenes; they are structurally spherical exhibiting various organic solvent solubility. The carbon cage structure of fullerenes comprises of fused ring system having pentagons and hexagons. The most common members of this family are C_{60} and C_{70}. There are 60 carbon atoms arranged in twelve pentagons and twenty hexagons in C_{60} molecule. C_{60} molecule is a truncated icosahedron structure which is a polygon with sixty vertices and thirty-two faces, of these faces twelve are pentagonal and twenty are hexagonal. A carbon atom is present on each vertex and their valences are satisfied by two single and one double bond (sp^2 hybridized carbon). The compound is aromatic, having many resonating structures. There are conjugated double bonds in fused rings. Fullerene cages are about 7.15Å in diameter and are one carbon atom thick. Fullerane ($C_{60}H_{60}$) is a fully saturated form of fullerene. There are various other forms of fullerenes as heterofullerens, norfulleres, homofullerene, secofullerene, etc.

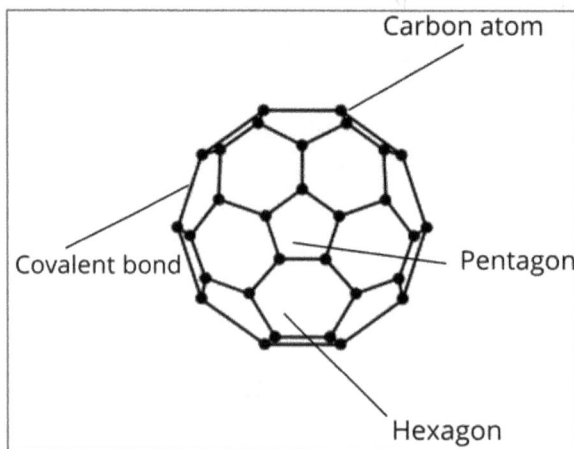

Figure 10.1 Structural details of fullerene.

The first proposal of buckyball was given by Eiji Osawa, Japan. He hypothesized that entire structure could exist and proposed that corannulene, a cyclopentane ring fused with five benzene rings was a part of football framework. In 1985, the group of scientists Richard Smalley, Robert Curl, James Heath, Sean O'Brien, and Harold Kroto synthesized the first fullerene molecule, buckminsterfullerene (C_{60}) at Rice University. They named the molecule in honor of the geodesic dome created by architect R. Buckminster Fuller having the same shape. Fullerenes are composed of fused hexagons and pentagons. C_{60} and C_{70} are most accessible members of this family. The high symmetry found in this molecule is an important property.

10.2 Classification of Fullerenes

C_{60} Fullerene: In C_{60} each carbon atom is attached to three other atoms using sp^2 orbitals with an electron in each orbital. The fourth valence electron of each carbon would be in an orbital p perpendicular to the spherical surface. In this way, the orbitals overlap forming a continuum of orbitals with electrons p inside and outside the sphere in the same way as it takes place in benzene with the six p electrons that give it the aromatic character. This means that we can consider fullerenes as an aromatic and stable sphere. The structure of the C_{60} is similar to that of a soccer ball. It is shaped like a truncated icosahedron with 60 vertices, in each of which is carbon. It has 32 faces, of which 12 are pentagons and the remaining 20 are hexagons, in addition each pentagon is surrounded by five hexagons so that two pentagons cannot be adjacent to each other, but the six links of each hexagon are alternately fused to three Pentagons and three hexagons.

C_{70} **Fullerene**: C_{70} fullerene has a closed, hollow fused-ring structure that looks like a cage. The overall structure of C_{70} fullerene resembles a rugby ball that is why they are often and commonly called 'bucky ball'. These 'bucky balls' are interconnected with 12 pentagons and 25 hexagons, in which, the atom of carbon is present at each Hexa- and pentagon's vertices, with a bond at each edge. A single atom of carbon is attached to the other three adjacent carbon atoms, bonded with single s and two p orbitals (sp^2 hybridization), which makes them stronger than the original ones, making them capable of making even stronger bonds. These formed molecules can undergo a variety of chemical reactions by accepting or donating an electron/s instantly.

Fullerenol: The derivatives of the fullerenes that are soluble in the water are called fullerenol. These molecules have the ability to remove free radicals. Fullerenols are commonly called as 'Radical Sponges', due to their characteristics of removing free radicals. They have great antioxidant properties due to the delocalized double pi-bond in the cages of fullerenes. They are mostly synthesized by the addition of the hydroxyl groups to C60 fullerene. Fullerenols possess a hollow-spherical shape as fullerenes.

Miscellaneous Fullerenes: The smallest fullerene is the C_{20}, which contains 20 pentagons and no hexagons. However, this type of structure has strong internal tensions because the shape of each carbon molecule is strongly non-planar. Other possible fullerenes are C_{28}, C_{32}, C_{44}, C_{50}, C_{58}, C_{70}, C_{76}, C_{84}, C_{240}, C_{540}, C_{960}, and many others.

10.3 Properties of Fullerenes

Some of the general properties of Fullerenes are summarized as under:

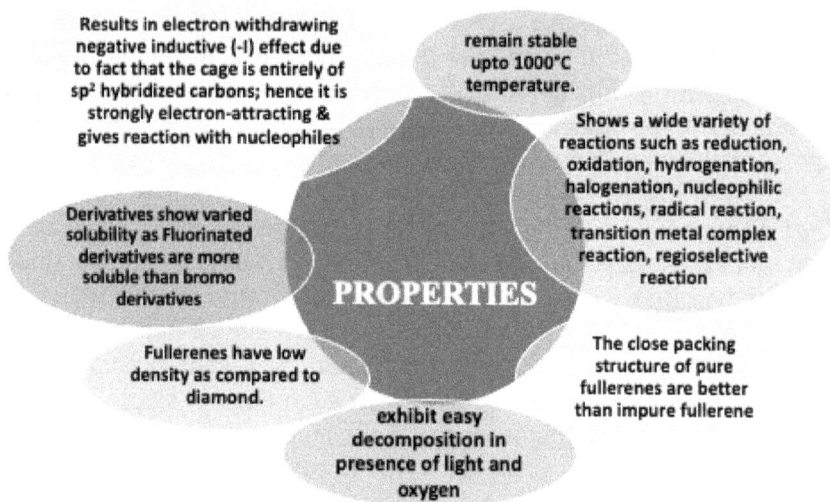

Figure 10.2 General properties of fullerenes.

The key factors that affect the **stability of fullerenes** are localization of electrons and ring strain. C_{60} has been found to be more aromatic and highly stable and the reason for this is reduced strain in ring as twelve pentagons are isolated from each other by hexagons in between. When there are minimum bond orders in five-membered rings it indicates that all double bonds are in six-membered rings and none in five-membered rings. The fusion of meta position of any two pentagonal rings with hexagonal ring results in an arrangement that permits minimization of bond orders in five-membered rings. It is established that as there occurs an increase in number of carbon atoms from sixty, stability starts decreasing.

10.4 Synthesis of Fullerenes

Pyrolysis

Pyrolysis of hydrocarbons (preferably aromatics) can also be used to get fullerenes. The pyrolysis of naphthalene at 1000° C in an argon steam was done first time to obtain fullerenes. The naphthalene skeleton is a small fragment of the C_{60} structure. The bowl-shaped corannulene was used as precursors for C_{60}, followed by naphthalene. Synthesis of fullerenes is also possible by laser pyrolysis which involves the use of benzene and acetylene as carbon source.

Arc Discharge Method

The advantage is simple construction of reactor, low cost which makes them attractive for synthesis. This procedure needs a welding transformer, a chamber connected to vacuum pump and graphite rods. The graphite electrodes are brought into close contact with each other, and an arc is struck in a helium or argon atmosphere. The soot is generated which is collected on a water-cooled surface. When extracted by toluene/benzene, this soot results in a reddish-brown solution being a mixture of fullerenes.

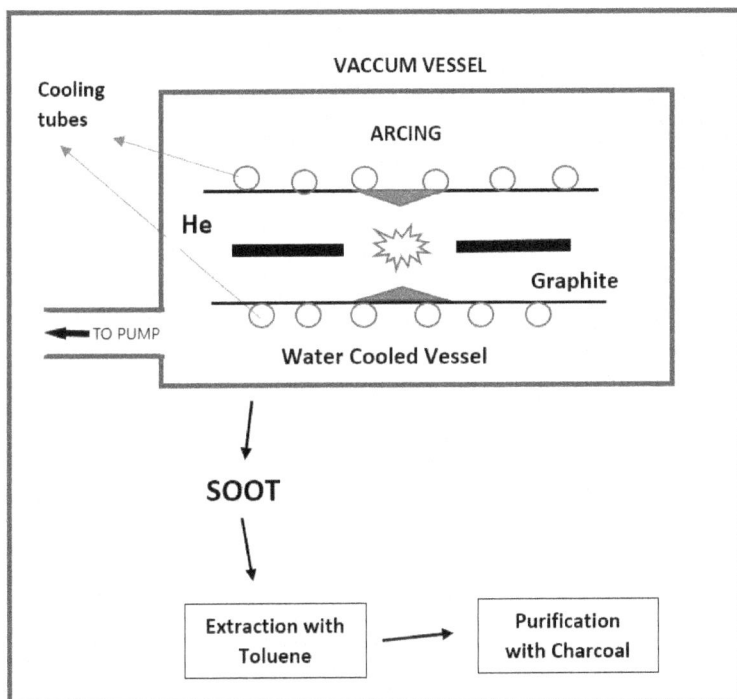

Figure 10.3 Block diagram of arc discharge method.

After extraction, the mixture is subjected to chromatographic separation. For example, C_{60} is collected by toluene as mobile phase; C_{70} can be collected by toluene/o-dichlorobenzene; higher fullerenes require HPLC for separation. Evaporation can also be used as a method for purification.

Laser Method

It is very powerful and useful technique for production of fullerene clusters. It involves laser ablation of graphite in helium atmosphere. Laser ablation is the process which involves removal of material from a solid (or occasionally liquid) surface with the help of irradiation using laser beam. As in this technique solvents are not used and operators are not exposed to chemicals, it is considered as environmentally friendly process. The material used in this process generally gets converted to plasma at higher laser flux. By carrying out this process at high temperature, this plasma cools more slowly and generated carbon clusters gets rearranged in stable fullerenes.

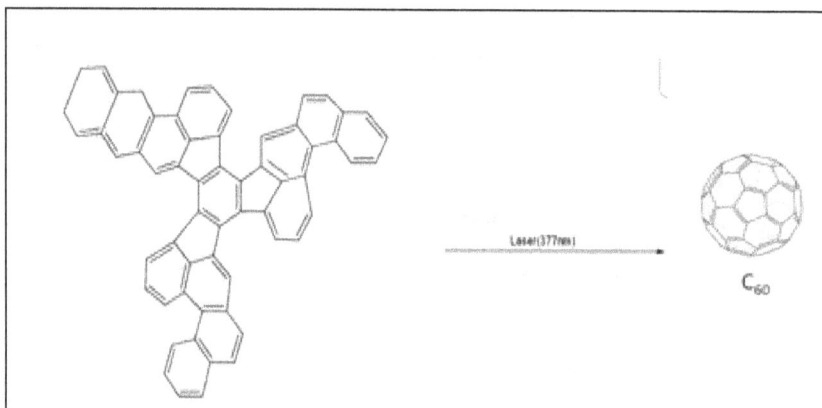

Figure 10.4 Synthesis of fullerenes by laser method.

10.5 Reactivity of Fullerenes

Fullerenes can be used in various organic reactions to form new compounds.

Nucleophilic Addition

As fullerenes have an inductive effect, they are easily attracted to nucleophiles. Various nucleophiles such as carbon, nitrogen, phosphorous and oxygen react with C_{60} to exhibit nucleophilic additions. The reaction proceeds via intermediate formation which is stabilized by various reactions. Nucleophilic addition reactions are of various types, such as addition of carbon nucleophiles, addition of cyanide, addition of hydroxides and alkoxides, addition of phosphorus, silicon, germanium nucleophiles and addition of macromolecular nucleophiles as in case of fullerene polymers.

1. Organolithium (R-Li) and Grignard compounds (R-Mg-X) react with C_{60} using alkyl, phenyl or alkenyl compounds to form primary intermediates. To obtain maximum yield, organolithium compounds as nucleophiles are used. The Bingel reaction is a very good example whose mechanism involves abstraction of proton by base to form carbanion or enolate which reacts with electron deficient fullerene double bond followed by displacement of bromine leading to intramolecular ring closure.

2. The negative shift of reduction potential results due to addition of alkyl nucleophile. To compensate this potential addition of strong electron withdrawing group is necessary such as addition of cyanid. C60 reacts with LiCN or NaCN to form monoadduct anion which can be quenched with various nucleophiles.

Cycloaddition Reaction

1. [4+2] Cycloaddition takes place under thermal conditions. The reaction of cyclopentadiene and C_{60} gives the monoadduct in comparatively high yield at room temperature. The anthracene cycloadduct is formed when excess of diene is refluxed in toluene.

2. [3+2] cycloaddition is shown when C_{60} reacts with diazomethane in toluene to obtain pyrazoline intermediate. Removal of N_2 form this intermediate results in two different bridged fullerenes.

Hydrogenation

As fullerenes have large ring strain, hence complete hydrogenation is not possible, yet, $C_{60}H_{18}$ and $C_{60}H_{36}$ are common examples of hydro fullerenes. The prolonged hydrogenation by direct reaction with hydrogen gas at high temperature results in cage formation which leads to instability of highly hydrogenated fullerenes. Fullerenes are easily hydrogenated by several methods. Hydrogenation of C_{60} was done with Lithium (Li) in liquid ammonia (Liq. NH_3). The major products of this are the isomers of $C_{60}H_{18}$ and $C_{60}H_{36}$.

10.6 Applications of Fullerenes

As antioxidants- Fullerenes can act as powerful antioxidants, which **react with free radicals** at high rate preventing cell damage/death. Fullerenes have been very promising in health and personal care sectors. It possesses a novel ability of selectively entering oxidation-damaged cerebral endothelial cells rather than normal endothelial cells and then protecting them from apoptosis. They have shown non-physiological applications in area where oxidation and radical processes are found to be destructive like food spoilage, plastic deterioration, metal corrosion, etc.

As additives for polymers- Fullerenes are chemically reactive to a great extent and hence on addition to polymers, they help in creating new **copolymers** with characteristic physical & mechanical properties. Along with helping in the formation of composites, they have also been of good worth in modifying physical properties and performance when added to polymers.

In drug delivery- Fullerenes form an efficient system of drug delivery. They serve as strong drug adsorbents. The fullerene derivatives which are water soluble and closely localized to mitochondria, have opened a new area of Fullerene Drug Delivery Systems. These water-soluble derivatives also inhibit the **HIV-1 protease**, which is an enzyme responsible for the development of the virus and are hence useful in fighting the HIV virus that causes AIDS.

As lubricants- Fullerenes offer composite coatings based on **inorganic fullerene-like material (IFLM)**. Nanosphere powders are being developed to reduce friction and improve wear resistance in places having rolling and sliding contacts as in ball bearing, gears, artificial joints, pumps, etc. When IFLM are fused into matrix, the particles allow independent control of friction and wear, with consistent performance on interacting surfaces in relative motion.

Miscellaneous Applications
1. Fullerene-based particles which are antibacterial, are coated on polystyrene surface to prevent biofilm formation of *Pseudomonas*.
2. As a superconductor on mixing with alkali metals.
3. Ferrocene is used in electronic, microelectronic and non-linear optical devices.
4. Fullerene can be used as a soft ferromagnet (C_{60} complexes with tetrakis(dimethylamino)ethylene (TDAE)).

10.7 Doped Fullerenes

Superconducting fullerenes based on C60 are fairly different from other organic superconductors. The building molecules are no longer manipulated hydrocarbons but pure carbon molecules. In addition to this, these molecules are no longer flat but bulky which gives rise to a three-dimensional, isotropic superconductor. The pure C_{60} grows in a fcc-lattice and is an insulator. By placing alkali atoms in the interstitials, the crystal becomes metallic and eventually superconducting at low temperatures.

Unfortunately, C_{60} crystals are not stable at ambient atmosphere. They are grown and investigated in closed capsules, limiting the measurement techniques possible. The highest transition temperature measured so far was T_C = 33 K for Cs_2RbC_{60}. The highest measured transition temperature of an organic superconductor was found in 1995 in Cs_3C_{60} pressurized with 15 kbar to be T_C = 40 K. Under pressure this compound shows a unique behavior. Usually, the highest TC is achieved with the lowest pressure necessary to drive the transition. Further increase of the pressure usually reduces the transition temperature. However, in Cs3C60 superconductivity sets in at very low pressures of several 100 bar, and the transition temperature keeps increasing with increasing pressure. This indicates a completely different mechanism then just broadening of the bandwidth. Examples of doped fullerenes are: K_3C_{60}, Rb_3C_{60}, K_2CsC_{60}, K_2RbC_{60}, K_5C_{60}, Sr_6C_{60}, $(NH_3)_4Na_2CsC_{60}$, $(NH_3)K_3C_{60}$, etc. The cage-like structure of the C_{60} molecule naturally offers two other possibilities for doping: inserting a foreign atom M inside the C_{60} cage and replacing one or several carbon atoms in the C_{60} cage with atoms having different electronic structure. In the first case the **endohedral super-molecule (1)** is obtained and in the second one the **composition is $C_{59}M$ (2)**.

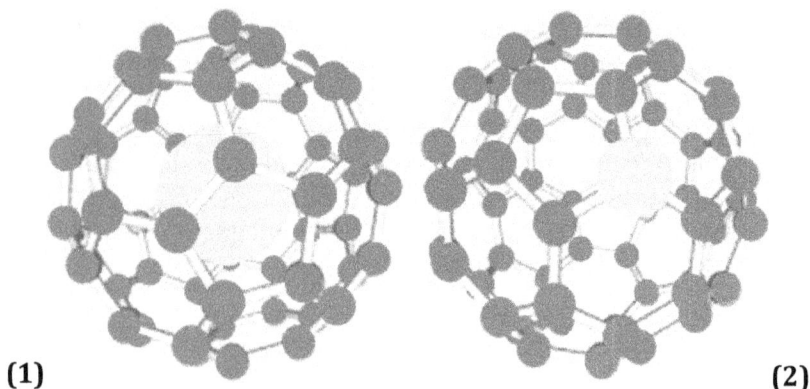

(1)   (2)

Figure 10.5 Possibilities of doping in fullerenes – (1) insertion and (2) replacement.

Endohedral (1) C60 molecules can be prepared by a Force Method, where highly accelerated ions of atoms are implanted to the C_{60} cage. The energy of the ions should be just enough to open up the cage and enter. The first collision should absorb and redistribute a good part of the initial kinetic energy so that the atoms does not escape the cage. Endohedral molecules may have **M = N, P, Li, Ca, Na, K, Rb** were produced this way in small quantities. Larger yield can be achieved by co-evaporation of the carbon and the metal in an arc discharge chamber (typical for fullerene production as described earlier).

On-ball doping has been achieved by replacing one carbon atom with a nitrogen atom obtaining $C_{59}N$ (azafullerenes). The replacement of N for C results in the following changes in the product molecule:

1. adds one extra electron to the cage
2. changes the structure locally
3. lowers the symmetry of the molecule
4. splits the degeneracy of the electronic orbitals.

Furthermore, this chemical substitution renders the molecule very reactive with a high electron affinity. At ambient temperature the $C_{59}N$ exists only in a dimerized phase. The intercalated azafullerene, $K_6C_{59}N$ has been investigated in detail. Its structure was found to be similar to K_6C_{60}, but, in contrast to K_6C_{60}, it was predicted to be a metal, although the prediction has not been tested experimentally.

References

1. Quo, Y., Karasawa, N., and Goddard, W. (1991) Prediction of fullerene packing in C60 and C70 crystals. *Nature*, **351**, 464–467.
2. Taylor, R., Hare, J. P., Abdul-Sada, A. K., and Kroto, H. W. (1990) Isolation, separation and characterisation of the fullerenes C60 and C70: the third form of carbon. *Journal of the Chemical Society, Chemical Communications*, **20**, 1423–1425.
3. Taylor, R., Langley, G., Kroto, H., and Walton, D. R. M. (1993) Formation of C60 by pyrolysis of naphthalene. *Nature*, **366**, 728–731.
4. Chatterjee, K., Parker, D. H., Wurz, P., Lykke, K. R., Gruen, D. M., and Stock, L. M. (1992) Fast one-step separation and purification of buckminsterfullerene, C60, from carbon soot. *Journal of Organic Chemistry*, **57**(11), 3253–3254.
5. Hu, Y., Sol-Daura, A., Yao, Y., Liu, X., Liu, S., Yu, A., Peng, P., Poblet, J. M., Rodrguez-Fortea, A., Echegoyen, L., and Li, F. (2020) Chemical reactions of cationic metallofullerenes: An alternative route for exohedral functionalization. *Chemistry - A European Journal*, **26**, 1748 – 1753
6. Bethune, D., Johnson, R., Salem, J., de Vries, M. S., and Yannoni, C. S. (1993) Atoms in carbon cages: the structure and properties of endohedral fullerenes. *Nature*, **366**, 123–128.
7. Kumar, A. (2017) Superconductivity of fullerenes. *Research Journal of Pharmaceuticals, Biology and Chemical Science*, **8**(3), 1045-1053.
8. Haddon, R. C., Hebard, A. F., Rosseinsky, M. J., Murphy, D. W., Duclos, S. J., Lyons, K. B., Miller, B., Rosmilla, J. M., Fleming, R. M., Kortan, A. R., Glarum, S. H., Makhija, A. V., Mullar, A. J., Eick, R. H., Zahurak, S. M., Tycko, R., Dabbah, G., and Thiel, F. A. (1991) Conducting films of C60 and C70 by alkali-metal doping. *Nature*, **350**, 320.
9. Song, L. W., Fredette, K. T., Chung, D. D. L., and Kao, Y. H. (1993) Superconductivity in interhalogen-doped fullerenes. *Solid State Communications*, **87**, 387.
10. Lieber, C. M., and Zhang, Z. (1994) Physical properties of metal doped fullerene superconductors. *Journal of Physics C: Solid State Physics*, **48**, 349-384.

11. Palstra, T. T. M., Zhou, O., Iwasa, Y., Sulewski, P. E., Fleming, R. M., and Zegarski, B. R. (1994) Electronic properties of metal doped fullerides. *Solid State Communications*, **92**, 71.
12. Kroto, H. W., Hearth, L. D., O'Brien, S. C., Curl, R. F., and Smalley, R. E. (1985) C60: Buckminsterfullerene. *Nature*, **318**, 162.
13. Rao, C. N. R., Seshadri, R., Govindaraj, A., and Sen, R. (1995) Fullerenes, nanotubes, onions and related carbon structures. *Materials Science and Engineering*, **15**, 209.
14. Forró, L., and Mihály, L. (2001) Electronic properties of doped fullerenes. *Reports on Progress in Physics*, **64**, 649.
15. Lia, G., and Sabirianov, R. F. (2008) Electronic and magnetic properties of endohedrally doped fullerene $Mn@C_{60}$: A total energy study. *The Journal of Chemical Physics*, **128**(7), 1376-1387.
16. Macovez, R. (2018) Physical properties of organic fullerene cocrystals. *Frontiers in Materials*, **4**, 46-58.

Chapter 11

Molecular Rectifiers and Transistors

Introduction

Charge transfer from one end to another can occur easily when molecules are incorporated between metal electrodes and hence produces current flow. Electron transfer or charge transfer reactions are the simplest reactions and occur in almost all-natural processes. They are closely dependent on the arrangement of molecular structure. Hence, the study of molecular structures and the electron transport through them remains significant.

Conventional electronics have traditionally been made from bulk materials. With bulk methods growing increasingly demanding and costly as they near inherent limits, the idea was born that the components could instead be built up atom by atom in a lab (bottom up) versus carving them out of bulk material (top down) (as already explained in Chapter 4). This is the idea behind molecular electronics, with the ultimate miniaturization being components contained in single molecules. Molecular electronics was coined by **Mark Ratner in 1974** and it involves the replacement of traditional electronic elements with one or more molecules. Simple molecular electronic devices comprise of molecules (organic in nature) placed between conducting electrodes.

In single-molecule electronics, the bulk material is replaced by single molecules. The molecules used should have properties that resemble traditional electronic components such as a **wire, transistor or rectifier**. Single-molecule electronics is an emerging field, and currently, the focus is on discovering molecules with interesting properties and on finding ways to obtain reliable and reproducible contacts between the molecular components and the bulk material of the electrodes.

11.1 Advantages and Disadvantages of Molecular Electronics

Molecular Electronics has certain **advantages** which make them more useful for instance, Morphologically, they are amorphous and polycrystalline, Charge carrier is a molecule, offer high mobility, can be processed at low temperatures and the synthetic flexibility is quite high. However, they have certain **limitations** as well like, Organic Molecular Materials are unstable, working on thin films does not give reliable electrical contacts, on exposure to air, water and UV light they show a drastic degradation in their electronic properties and they cannot be used in high frequency devices.

11.2 Rectifiers

A rectifier is an electrical device responsible for converting alternating current (**AC**), periodically reversing its direction, to direct current (**DC**). It is similar to a one-way valve that permits an electrical current to flow only in one direction. This process is known as **rectification**.

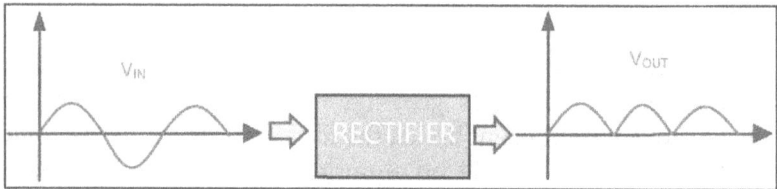

Figure 11.1 Process of rectification.

Apart from the primary application of the rectifier to convert AC power to DC power, they are used inside the power supplies of almost all electronic equipment where they are placed in series following the transformer, a smoothing filter, and possibly a voltage regulator. **Bridge rectifiers** are widely used for large appliances, where they are capable of converting high AC voltage to low DC voltage. The use of a **half-wave rectifier** can help us achieve the desired dc voltage by using a step-down or step-up transformers. **Full-wave rectifiers** are even used for powering up the motor and led, which works on DC voltage. A half-wave rectifier is used in **soldering** iron types of circuit and is also used in mosquito repellent to drive the lead for the fumes. In electric welding, bridge rectifier circuits are used to supply steady and polarized DC voltage. A half-wave rectifier is used in **amplitude modulation radio** as a detector because the output consists of an audio signal. Due to the less intensity of the current, it is of very little use to the more complex rectifier. A half-wave rectifier is used in firing circuits and pulse generating circuits. For demodulating the amplitude of a modulated signal, a half-wave rectifier is used. In a radio signal, to detect the amplitude of a modulating signal, a full-wave bridge rectifier is used. It is used in **voltage multiplier**. For the purpose of the voltage multiplier, a half-wave rectifier is used.

A rectifier can take several physical forms such as solid-state diodes, vacuum tube diodes, mercury-arc valves, silicon-controlled rectifiers, and various other silicon-based semiconductor switches.

11.3 Molecular Rectifiers

Rectifier/Diode based logic circuits are well known for building logic families by using the rectifying diodes at the input and connecting a resistor between the supply or the ground. A **molecular diode/rectifier** also contains two terminals and functions like a semiconductor **p-n junction** (an interface having positive and negative sides and allows current to flow through the interface only in one direction) and has electronic states: one is highly conductive state (**ON**) and other is a less conductive state (**OFF**).

The building of molecular diodes/rectifiers has been possible because of the work of Aviram and Ratner in 1974. They have suggested that electron donating components make conjugated molecular groups having a **large electron density (N-type)** and electron withdrawing components make conjugated molecular groups having **low electron density (P-type)**. According to them, a **non-centrosymmetric molecule** (does not contain an inversion center as one of its symmetry elements) having appropriate donor and acceptor moieties linked with a s-bridge and connected with suitable electrodes will conduct current only in one direction - acting as a rectifier.

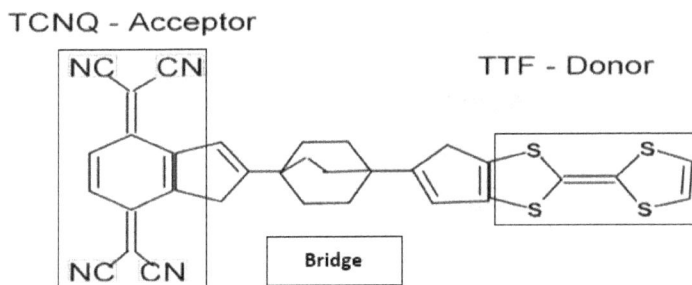

Figure 11.2 Layout of donor-bridge-acceptor molecule (example 1).

Thus, in this **donor-bridge-acceptor** molecule, the lowest unoccupied molecular orbital (**LUMO**) and highest occupied molecular orbital (**HOMO**) will be aligned in a way that conduction of electrons is only in one direction, hence functioning as a molecular diode/rectifier. Elaborating it, the **e⁻ donor elements** (n-type) increase the pi density and lower ionization potential (i.e., increase HOMO) while the **e⁻ acceptor elements** (p-type) decrease the pi density and higher ionization potential (i.e., decrease LUMO).

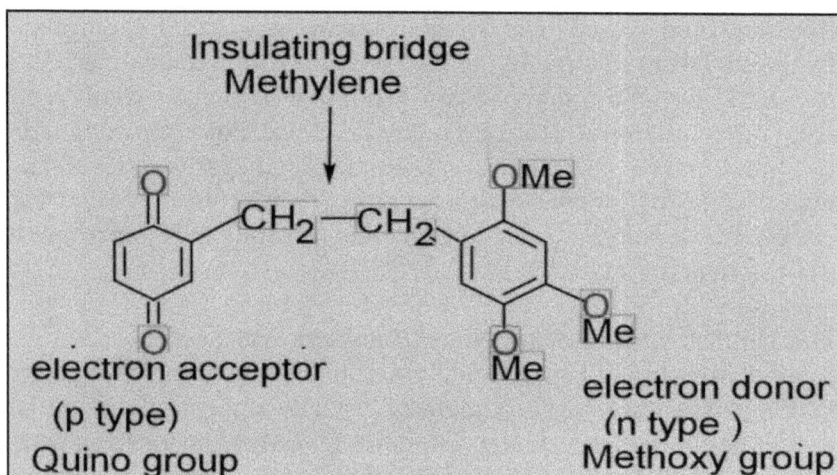

Figure 11.3 Layout of donor-bridge-acceptor molecule (example 2).

This diode is based on a molecular conducting wire consisting of two identical sections (S1, S2) separated by an insulating group R. Structure of molecular diode integrally embedded in a poly-phenylene based molecular conducting wire.

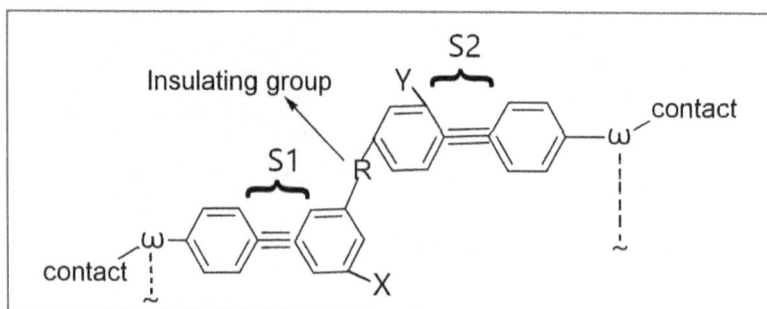

Figure 11.4 Structure of molecular diode.

Section S1 is doped by at least one electron donating group (X) and section S2 is doped by at least one electron withdrawing group (Y). The insulating group can be incorporated into the molecular wire by bonding a saturated aliphatic group (no pi-orbitals). To adjust the voltage-drop across R, multiple donor/acceptor sites can be incorporated. The single molecule ends are connected to the contact electrodes e.g., gold.

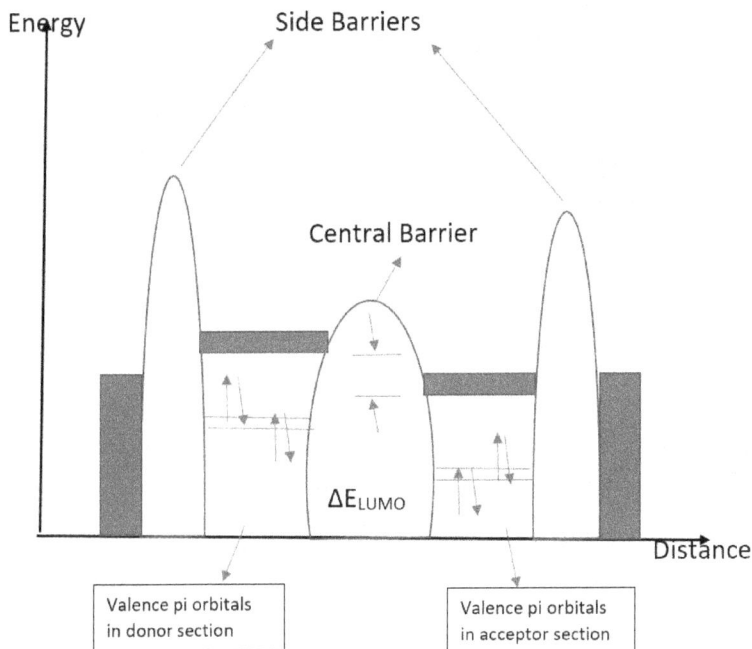

(a)

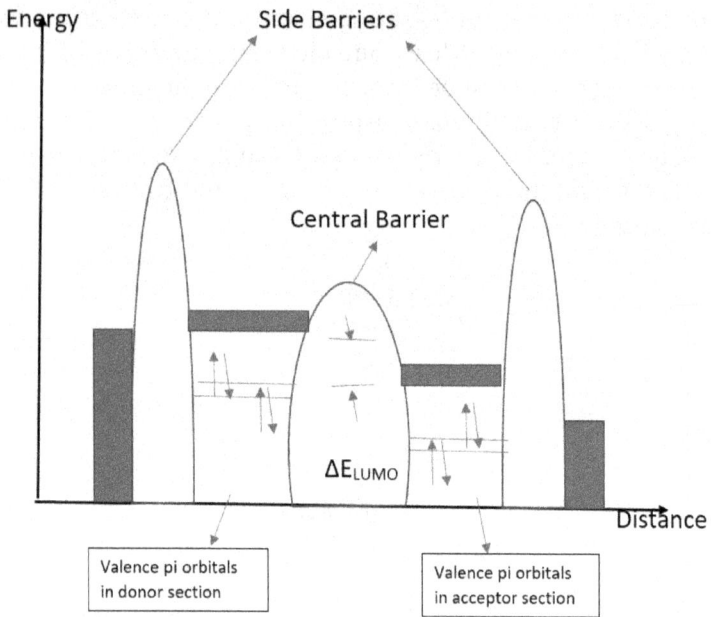

(b)

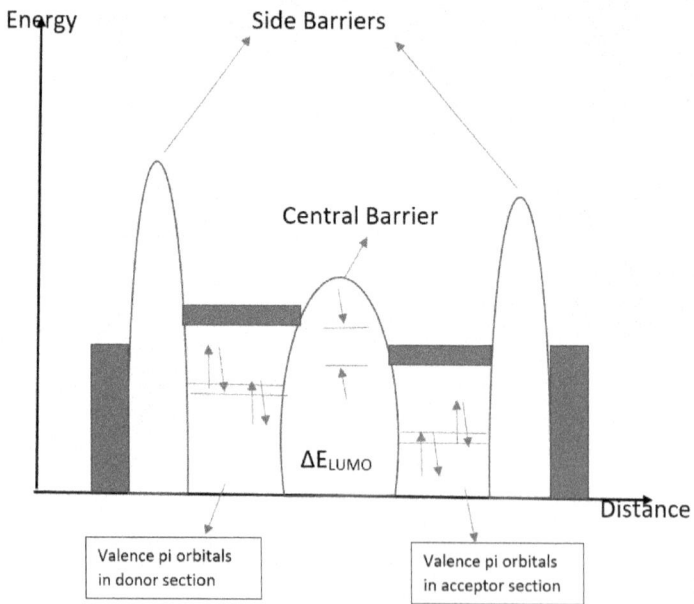

(c)

Figure 11.5 Energy barrier in forward (a), reverse (b) and zero (c) bias.

It can be seen that there are three potential barriers - one corresponding to the insulating group (middle barrier) and two corresponding to the contact between the molecule and the electrode (left and right barriers).

These potential barriers provide the required isolation between various parts of the structure. The occupied energy levels in the metal contacts and the Fermi energy level EF are also shown. On the left of the central barrier all the pi-type energy levels (HOMO as well as LUMO) are elevated due to the presence of the electron donating group X and similarly on the right of the central barrier the energy levels are lowered due to the presence of the electron withdrawing group Y.

This causes a built-in potential to develop across the barrier represented by the energy difference DELUMO. For current to flow electrons must overcome the potential barrier from electron acceptor doped section (S2) to electron donor doped section (S1) and this forms the basis for the formation of the **mono-molecular rectifying diode**. Here, electrons are induced to flow by tunneling through the three potential barriers from right to left causing a forward current flow from left to right.

11.4 Molecular Transistors

The term molecular transistors refer to switching circuits constructed from an individual molecule. Transistors form the basis of electronic circuits and have been the main constituent of the digital revolution. It is an electronic component consisting of three terminals. On application of a voltage or current to one pair of the terminals, we can regulate the current through another pair of terminals. The transistor acts as an amplifier because output power becomes higher than the input power after going through it. Another application of transistor is that of an electrical switch, in which the voltage on one of the terminals regulates the current between other two terminals. Also, they have attracted much attention, and in 2000 the first **single-molecule transistor** was fabricated.

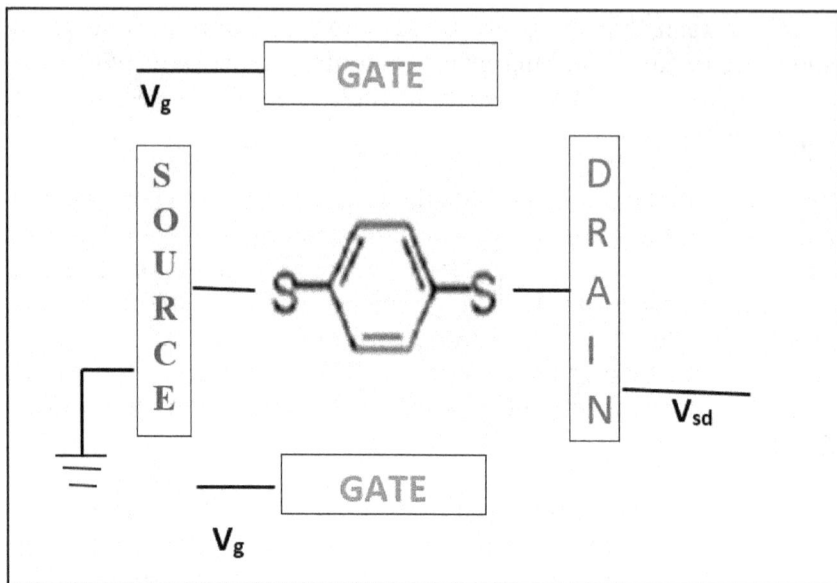

Figure 11.6 Block diagram of single-molecule transistor.

In that device, the current between the source and drain was regulated using a voltage applied on the gate electrode, which was achieved using a **single molecule** bridging the source and drain, below which the gate electrode was located. While a silicon semiconductor transistor is mostly used as an amplifier and a switch with a high-power gain, on the single molecule scale, the added value of the third electrode depends mainly on the supported spectroscopic information. By applying a voltage on the gate, we can change the electrostatic potential of the molecule. Energy shifts of the molecular transport level are then induced, from which supported information about the molecule can be obtained.

11.5 Benefits of Molecular Transistors Over Conventional Transistors

Since the development of electronic circuits, the stress has been laid on downsizing of these devices but in doing so, after a certain point, the resistance increases manifold thereby decreasing the efficiency of the device. The conventional devices also lag behind in their horizontal topography as they are prepared using UV-lithography/Electron-Beam lithography. However, vertical topography is more logical.

There can be applied a control on layer thickness up to the lattice parameter length scale, during the use of evaporation technique. Also, unipolar devices could be an improvement as bipolar transistors exhibit a storage time of minority carriers leading to speed limitations. The transport mechanism now used should be high velocity carriers instead of the conventional diffusion mechanism. Thus, the tunneling hot carrier transistor uses vertical topography, unipolarity and high velocity carriers in their device structure clearing all vices of conventional transistors.

The THCT (tunneling hot carrier transistor) launches electrons to speed up the tunneling electrons from the emitter to the base. It consists of a set of three electrodes having thin insulator between electrodes, which makes a combination of MIMIM i.e., metal-insulator-metal-insulator-metal transistor. This increases the frequency performance of the device and can also be used as spin transport devices. Making of logic circuits with very few components can also be seen as an advantage of THCT since it has a feature of negative differential resistance.

The important part of MIMIM are the insulating layers which different purposes:

1. 1st Insulating layer – electron injector and accelerator
2. 2nd Insulating layer - electron selector sensitive to the energy of the incident electrons

Suppose two metals are connected with an energy barrier in between. If the barrier is thin enough the wave functions of the electrons at the interface extend into the other metal. So, the electrons have a certain probability to end up at the other side of the barrier, which is called tunneling. When a bias is applied to the metals a net tunneling current starts flowing.

In a MIMIM transistor, this tunneling principle is used to inject electrons from the emitter into the base. The tunnel junction generates a beam of high energy electrons having very high velocity as they are accelerated by a field when they tunnel through the barrier. A bias leads these electrons to the collector and with this last element, a three terminal device is completed.

References

1. Dyck, C., and Ratner, M. (2015). Molecular rectifiers: A new design based on asymmetric anchoring moieties. *Nano Letters*, **15**(3), 1577-1584.
2. Mathew, P. T., and Fang, F. (2018) Advances in molecular electronics: A brief review. *Engineering*, **4**(6), 760-771.
3. Ratner, M. (2013) A brief history of molecular electronics. *Nature Nanotechnology*, **8**, 378-381.
4. Heath, J. R., and Ratner, M.A. (2003) Molecular electronics. *Physics Today*, **56**(5), 43-49.
5. McCreery, R. L. (2004) Molecular electronic junctions. *Chemistry of Materials*, **16**, 477-4496.
6. Kushmerick, J. (2009) Molecular transistors scrutinized. *Nature*, **462,** 994-995.
7. Yu, H., Luo, Y., Beverly, K., Stoddart, J. F., Tseng, H. R., and Heath, J. R. (2003) The molecule-electrode interface in single-molecule transistors. *Angewandte Chemie (International Edition)*, **42,** 5706-5711.
8. Ghosh, A. W., Rakshit, T., and Datta, S. (2004) Gating of a molecular transistor: electrostatic and conformational. *Nano Letters*, **4**, 565-568.
9. Reed, M. A., Zhou, C., Muller, C. J., Burgin, T. P., and Tour, J. M. (1997) Conductance of a molecular junction. *Science*, **278**, 252-254.
10. Sotthewes, K., Geskin, V., Heimbuch, R., Kumar, A., and Zandvliet, H. J. W. (2014) Research update: molecular electronics: the single-molecule switch and transistor. *APL Materials*, **2**(1), 010701.
11. Aviram, A., and Ratner, M. A. (1974) Molecular rectifiers. *Chemical Physics Letters*, **29**(2), 277-283.
12. Martin, A. S., Sambles, J. R., and Ashwell, G. J. (1993) Molecular rectifier. *Physical Review Letters*, **70**(2), 218-221.
13. Tao, N. J. (2007) Electron transport in molecular junctions. *Nature Nanotechnology*, **1**(3), 173-181.
14. Ellenbogen, J. C. (1998) A Brief Overview of Nanoelectronic Devices. *Microelectronics Applications Conference*, Arlington, USA.
15. Kumar, M. J. (2007) Molecular diodes and applications. *Recent Patents on Nanotechnology*, **1**, 51-57.

Chapter 12

Artificial Photosynthetic Devices

Introduction

Photosynthesis is the process by which plants use sunlight, water, and carbon dioxide to create oxygen and energy in the form of sugar.

$$6CO_2 + 12H_2O + \text{Light Energy} \longrightarrow C_6H_{12}O_6 + 6O_2 + 6H_2O$$

12.1 Stages of Natural Photosynthesis

There are **four** identified stages of photosynthesis:

Capturing of Sunlight - This involves the absorption and concentration of electromagnetic radiation by chlorophyll. These molecules are packed together in protein complexes or organelles and serve to concentrate the captured energy in reaction centers.

Segregation of Charge - At reaction center, charge separation takes place: a chlorophyll molecule releases an electron, leaving a positively charged 'hole'. In this way, energy from sunlight is used to separate positive and negative charges from each other.

Breakdown of Water - In the third step, many positive charges are collected and used to break water molecules into hydrogen-ions and oxygen. This is carried out in a separate compartment away from the charge separation stage: far enough to prevent loss of charge upon arrival of the next photon, and close enough to allow for efficient buildup of charge for catalysis.

Preparation of Food/Fuel - Electrons from the charge separation step are transferred via cytochrome and small mobile electron carriers to another protein complex. Here, more energy is added using photons from sunlight and the electrons are then used in a chemical reaction that finally produces carbohydrates.

It consists of two major processes:

    a. Light Dependent Reaction, where plants use sunlight energy to split water and make oxygen.

    b. Dark Reaction (Calvin cycle), where fixation of carbon di oxide resulting in formation of sugar molecule from carbon di oxide and water.

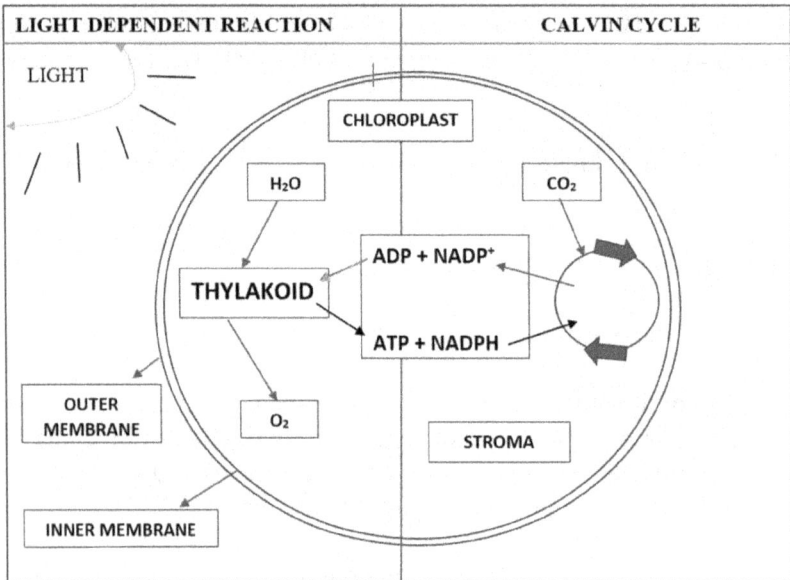

Figure 12.1 Procedure of natural photosynthesis.

12.2 Artificial Photosynthesis

Artificial photosynthesis is described as a chemical process that **impersonates** the natural process of photosynthesis to convert sunlight, water, and carbon dioxide into carbohydrates and oxygen. The term artificial photosynthesis is commonly used to refer to any **process** for capturing and storing the energy from sunlight in the chemical bonds of a food/fuel (called a solar fuel). The artificial photosynthetic device consists of four components:

a) Photosensitizer, which harvests light and transfers electrons to the hydrogen catalyst and itself getting oxidized.

b) Reaction Hub, it is a center which maintains the electron gradient by gaining and loosing of electrons.
c) Catalytic Site 1, which induces the water splitting catalyst to give electrons to the gradient, hence referred to as Donor. The oxidized donor is able to bring about oxidation of water.
d) Catalytic Site 2, where liberated protons and electrons be used to reduce carbon dioxide to storable fuels such as methanol.

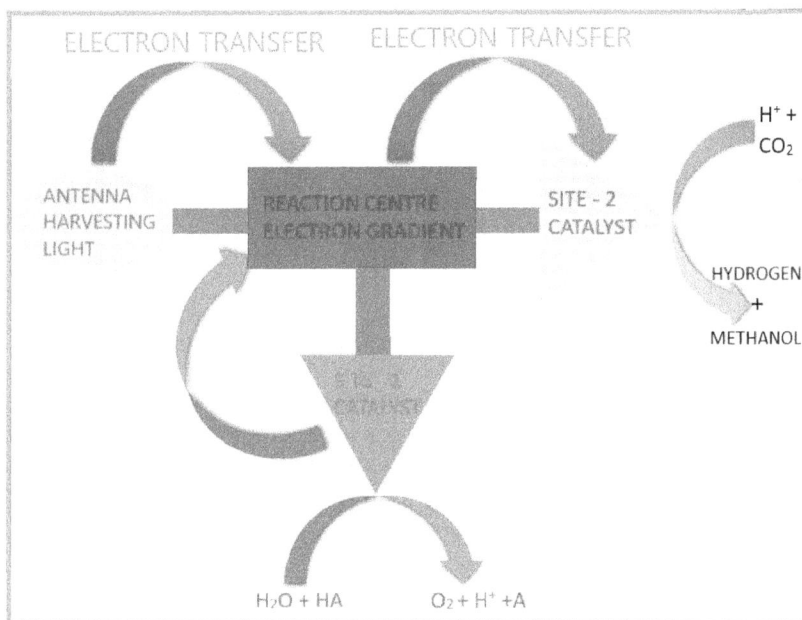

Figure 12.2 Block diagram of artificial photosynthesis.

The major task at hand is to obtain the best catalyst for executing various procedures efficiently to bring out best results.

Catalyst for Hydrogen

The simplest solar fuel to form is Hydrogen, as it requires the transference of two electrons to two protons. It is formed stepwise, with the hepl of an intermediate hydride anion:

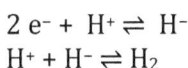

$2 e^- + H^+ \rightleftharpoons H^-$
$H^+ + H^- \rightleftharpoons H_2$

The catalyst for converting proton-to-hydrogen available in nature are **hydrogenases**. They are enzymes working both ways i.e., either reduce protons to hydrogen molecule or oxidizing hydrogen molecule to protons and electrons. The functionally identical of nickel-iron and iron-iron hydrogenases found in nature are structural H-cluster models, a dirhodium photocatalyst, and cobalt catalysts.

Catalyst for Oxidizing Water

Oxidation of water is a more complex chemical reaction than reduction of proton. In nature, the oxygen-evolving complex performs this reaction by accumulating reducing equivalents (electrons) in a manganese-calcium cluster within photosystem II (PS II), then delivering them to water molecules, with the resulting production of molecular oxygen and protons:

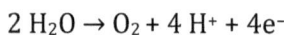

$$2\,H_2O \rightarrow O_2 + 4\,H^+ + 4e^-$$

The above reaction, if conducted without a catalyst, is highly endothermic and requires a temperature of approximately 2500 K. The oxygen evolving catalytic complex is a **clustered structure** containing four manganese ions and one calcium ions. Another example is of ruthenium complexes which is capable of water oxidation in the presence of light. In this case, the **ruthenium complex** acts as both photosensitizer and catalyst. Other metal oxides also serve as catalyst for water oxidation, for example, ruthenium(IV) oxide (RuO_2), iridium(IV) oxide (IrO_2), cobalt oxides (including nickel-doped Co_3O_4), manganese oxide (MnO_2, Mn_2O_3 and a mixture of Mn_2O_3 with $CaMn_2O_4$).

Recently **metal-organic framework** (MOF)-based materials have been shown to be a highly promising candidate for water oxidation with first row transition metals.

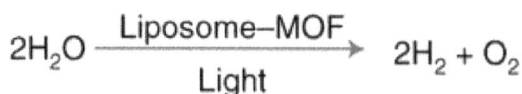

$$2H_2O \xrightarrow[\text{Light}]{\text{Liposome–MOF}} 2H_2 + O_2$$

$$\text{Quantum yield} = (1.5 \pm 1)\%$$

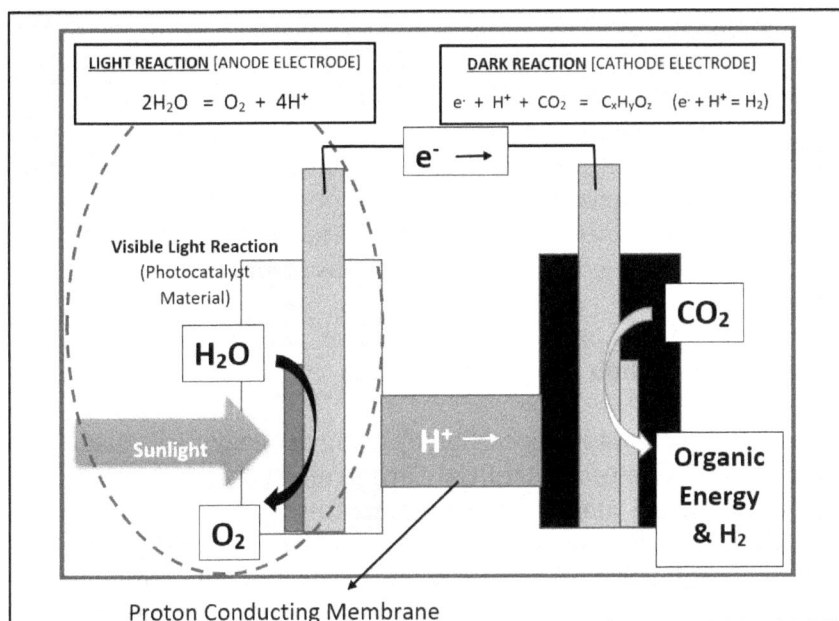

Figure 12.3 Setup for artificial photosynthesis.

Capturing of Light via Photosensitizers

Natural photosynthesis uses chlorophylls for absorption of a broad part of the visible spectrum. Artificial photosynthesis can either do so with the help of one type of pigment, having a broad absorption rang,e or a mixture of several pigments for the same purpose. Tris(bi-pyridine)ruthenium ion, $[Ru(bipy)3]^{2+}$, acts as a photosensitizer.

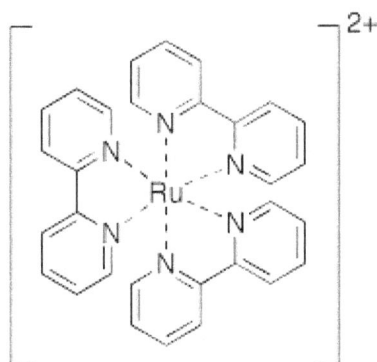

Ruthenium polypyridine complexes, in particular tris(bipyridine)ruthenium(II) and its derivatives, have been extensively used due to their efficient visible light absorption for hydrogen photoproduction. Other noble metal-containing complexes used include ones with platinum, rhodium and iridium. Pyrrole rings such as porphyrins have also been used in coating nanomaterials or semiconductors for both homogeneous and heterogeneous catalysis. Zeolite framework as a host for organic dyes have also been proposed for the purpose of photosensitizer.

Catalyst for Carbon Dioxide Reduction

In nature, green plants fix carbon by using the enzyme RuBisCO as a part of the Calvin cycle. Artificial Carbon di oxide reduction for producing food/fuel aims mostly at generating reduced carbon compounds from atmospheric CO_2. Some **transition metal poly-phosphine complexes** have been developed for this purpose; however, they usually require previous concentration of CO_2 before use, and carriers that are both stable in aerobic conditions and able to concentrate CO_2 at atmospheric concentrations. The simplest product from CO_2 reduction is carbon monoxide (CO), but for fuel development, further reduction is needed, and a key step also needing further development is the transfer of hydride anions to CO.

References

1. Hu, H., Wang, Z., Cao, L., Zeng, L. Zhang, C. Lin, W., and Wang, C. (2021) Metal–organic frameworks embedded in a liposome facilitate overall photocatalytic water splitting. *Nature Chemistry*, **13**, 358–366.
2. Bensaid, S., Centi, G., Garrone, E., Perathoner, S., and Saracco, G. (2012) Towards artificial leaves for solar hydrogen and fuels from carbon dioxide. *ChemSusChem*, **5**(3), 500–521.
3. Morris, A. J., Meyer, G. J., and Fujita E. (2009) Molecular approaches to the photocatalytic reduction of carbon dioxide for solar fuels. *Accounts of Chemical Research*, **42**(12), 1983–1994.
4. Umena Y., Kawakami K., Shen J. R., and Kamiya N. (2011) Crystal structure of oxygen-evolving photosystem II at a resolution of 1.9 angstrom. *Nature*, **473**, 55-65.

5. Artero, V., Chavarot-Kerlidou, M., and Fontecave, M. (2011) Splitting water with cobalt. *Angewandte Chemie International Edition*, **50**(32), 7238–7266.

6. Duan, L. L., Tong, L. P., Xu, Y. H., and Sun, L. C. (2011) Visible light-driven water oxidation-from Molecular catalysts to photoelectrochemical cells. *Energy and Environment Science*, **4**(9), 3296–3313.

7. Lubitz, W., Reijerse, E. J., and Messinger, J. (2008) Solar water-splitting into H_2 and O_2: design Principles of photosystem II and hydrogenases. *Energy and Environment Science*, **1**(1), 15–31.

8. Herrero, C., Quaranta, A., Leibl, W., Rutherford, A. W., and Aukauloo, A. (2011) Artificial photosynthetic systems: using light and water to provide electrons and protons for the synthesis of a fuel. *Energy and Environment Science*, **4**(7), 2353–2365.

9. Andreiadis, E. S., Chavarot-Kerlidou, M., Fontecave, M., and Artero, V. (2011) Artificial photosynthesis: From molecular catalysts for light-driven water splitting to photoelectrochemical cells. *Photochemistry and Photobiology*, **87**(5), 946–964.

10. Nocera D. G. (2012) The artificial leaf. *Accounts of Chemical Research*, **45**(5), 767–776.

11. Badura, A., Kothe, T., Schuhmann, W., and Rogner. M. (2011) Wiring photosynthetic enzymes to electrodes. *Energy and Environment Science*, **4**(9), 3263–3274.

Chapter 13

Optical Storage Memory

Introduction

A storage device is used in the computers to store the data and provides one of the core functions of the modern computer. The electronic storage medium that uses low-power laser beams to record and retrieve digital (binary) data is termed as **optical storage**. In optical-storage technology, the working consists of a laser beam that encodes digital data onto an optical, or laser, disk in the form of tiny pits arranged in a spiral track on the disk's surface. A low-power laser scanner reads these pits, with variations in the intensity of reflected light from the pits being converted into electric signals.

The optical storage has been **classified** into four types viz., primary storage, secondary storage, tertiary storage and offline storage.

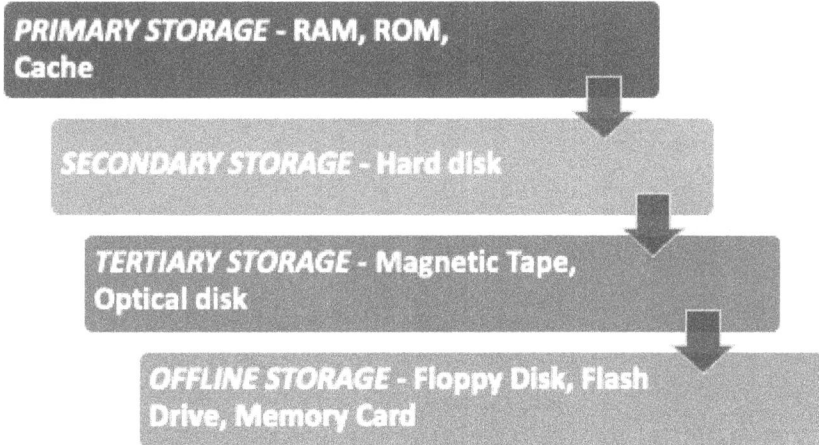

PRIMARY STORAGE - RAM, ROM, Cache

SECONDARY STORAGE - Hard disk

TERTIARY STORAGE - Magnetic Tape, Optical disk

OFFLINE STORAGE - Floppy Disk, Flash Drive, Memory Card

13.1 Primary Optical Storage

Primary optical storage is also referred to as main memory and it is called so because it is directly or indirectly connected to the central processing unit via a memory bus. The central processing unit continuously reads instructions stored there and executes them as required. Examples are RAM, ROM and Cache. **RAM (1)** is called **random access memory** because any of the data available in RAM can be accessed immediately as fast as any of the other data. RAM is divided into two types: **DRAM** (dynamic random-access memory) and **SRAM** (static random-access memory). **ROM** is called read only memory and is used as the computer boots/starts up. Small programs called firmware are often stored in ROM chips on hardware devices (like a BIOS chip), and they contain instructions the computer can use in performing some of the most basic operations required to operate hardware devices. Hence, when the system is restored to factory settings, then only these basic programs run, and the system starts working as it was at the initial stage. **Cache (2)** is a high-speed access area that can be either a reserved section of main memory or a storage device. It is used to speed up the memory retrieval process. Every time, as a command is given, the central processing unit requests for data from the memory unit. The cache stores frequently used data and data around it to make this retrieval process faster. Since the cost of cache is higher, the size of cache in central processing unit is much smaller as compared to main memory.

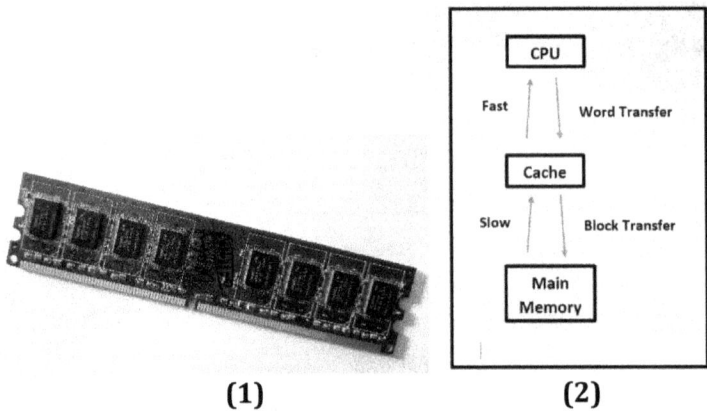

(1)   (2)

Figure 13.1 Primary optical storage: (1) RAM and (2) Cache.

13.2 Secondary Optical Storage

Secondary optical storage is not directly accessed by the central processing unit. Computer usually uses its input/output channels to access secondary storage and transfers the desired data using

116

intermediate area in primary storage. Example: __hard disk__. The hard disk drive is the main, and usually largest, data storage device in a computer. It can store anywhere from gigabytes to terabytes. The speed of the Hard disk is the speed at which content can be read and written on a hard disk. Disk access time is measured in milliseconds. There are two types of hard disks: **internal hard disk**, which has zero portability and bigger in size and **external hard disk**, which is portable and smaller in size.

(1)

(2)

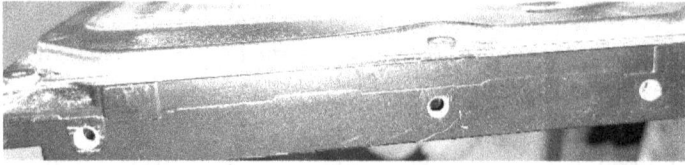

(3)

Figure 13.2 Secondary optical storage: (1) external hard disk, (2) internal hard disk (rear view) and (3) internal hard disk (lateral view).

13.3 Tertiary Optical Storage

Tertiary optical storage is an exhaustive computer storage system which is usually very slow, so it is used to archive data that is not accessed frequently. It is mainly useful for extraordinarily large data stores. Examples: **magnetic tape,** which is a magnetically coated strip of plastic on which data can be encoded. Tapes for computers are similar to tapes used to store music. Tape is much less expensive than other storage mediums but commonly a much slower solution that is commonly used for backup.

Figure 13.3 Magnetic tape.

Optical disc is a storage medium where information is stored as a series of lands, or flat areas, and pits. Optical storage provides greater memory capacity than magnetic storage because laser beams can be controlled and focused much more precisely than can tiny magnetic heads, thereby enabling the condensation of data into a much smaller space. Optical disks are also inexpensive to make and require mini-mum maintenance. The most common types of optical media are: **blu-ray (BD) (1), compact disc (CD) (2) and digital versatile disc (DVD) (3).**

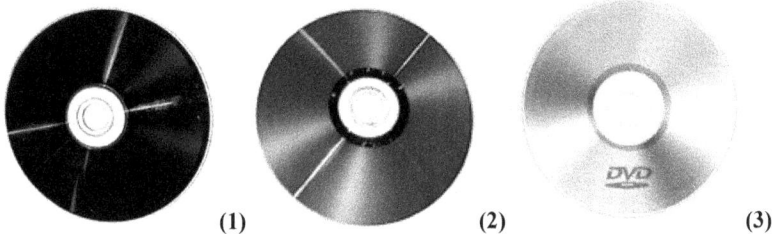

(1) (2) (3)

Figure 13.4 Optical disks: (1) blu-ray disk, (2) compact disk and (3) digital versatile disk.

There are different categories of Optical disc like:

1. **Read-only memory (ROM) disks-** They are used for large scale distribution of standard program and data files. These are mass-pro-duced by mechanical pressing from a master disc. The information is actually stored as physical indentations on the surface of the CD, like the audio CD.

2. **Write-once read-many (WORM) disks-** Some optical disks can be recorded once. The information stored on the disk cannot be changed or erased. Generally, the disk has a thin effective film deposited on the surface. There is a write protection facility which affords the as-surance that the data cannot be tampered with, once it is written to the device.

3. **Re-writeable, write-many read-many (WMRM) disks-** They al-low information to be recorded and erased many times. Usually, there is a separate erase cycle although this may be transparent to the user. They are also called direct-read-after-write disks.

119

The **working** of an optical disc includes a laser assembly that reads the spinning disc, converting lands and pits into sequences of electric signals. To store data, the disk's metal surface is covered with tiny dents (pits) and flat spots (lands), which cause light to be reflected differently. When the beam hits a land, it is reflected onto a photodiode, which produces an electric signal. Laser beams are scattered by pits, so no signal is generated. When an optical drive shines light into a pit, the light cannot be reflected back. This represents a bit value of 0 (off). A land reflects light back to its source, representing a bit value of 1 (on).

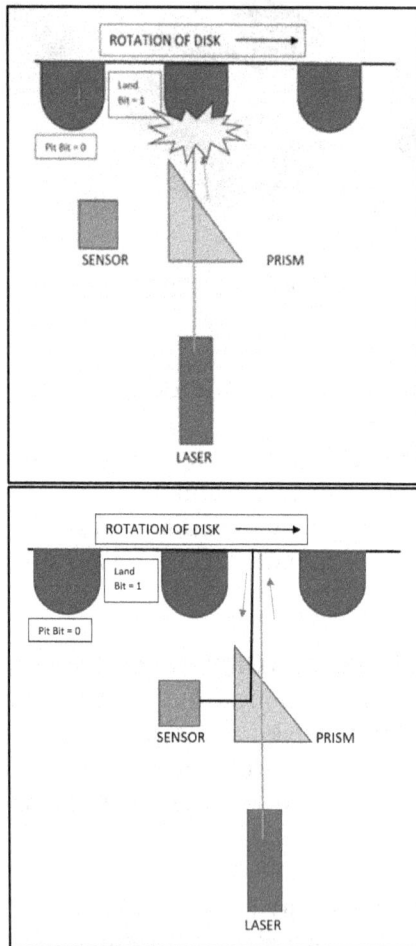

Figure 13.5 Working of an optical disk.

13.4 Offline Storage

Offline storage is also called removable storage and it is a medium or device used for data storage which is not under the control of a processing unit. It must be inserted or connected by a human operator before a computer can access it again. Examples: **flash drive (1), and memory card (2)**. A small, portable flash memory card that plugs into a computer's universal serial bus (USB) port and functions as a portable hard drive. Flash drives are available in various sizes and are a convenient way to transfer and store information. Memory Card is an electronic storage disk commonly used in digital cameras, MP3 players, mobile phones, and other small portable devices. Memory cards are usually read by connecting the device containing the card to a computer, or by using a USB card reader.

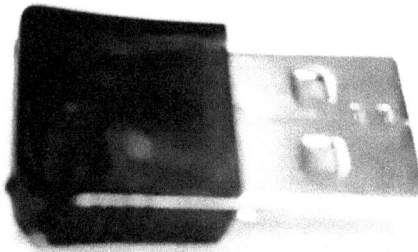

(1)

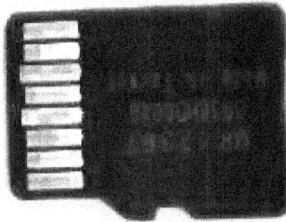

(2)

Figure 13.6 Offline storage: (1) flash drive and (2) memory card.

13.5 Principle of Optical Storage Memory

The recording of optical memory is based on the principle of **holography** and it is for this reason that the optical memory is also termed as holographic memory system. For data storage, optical memory is a promising since it is a true three-dimensional storage system, there is no limitation of accessing data sequentially, but it can be accessed by an entire page at a time and the mechanical motion has been minimum as there are very few moving parts. A photosensitive material is used by optical memory which records interference patterns of a reference beam and a signal beam of coherent light. The signal beam is reflected off of an object or it contains data in the form of light and dark areas. If a beam of light is applied to the material identical to the reference beam, then the nature of the photosensitive material should be such that the recorded interference pattern can be reproduced. The resulting light that is transmitted through the medium will take on the recorded interference pattern and will be collected on a laser detector array that encompasses the entire surface of the holographic medium. By changing the angle or the wavelength of the incident light, many holograms can be recorded in the same space. This is the way to access an entire page of data.

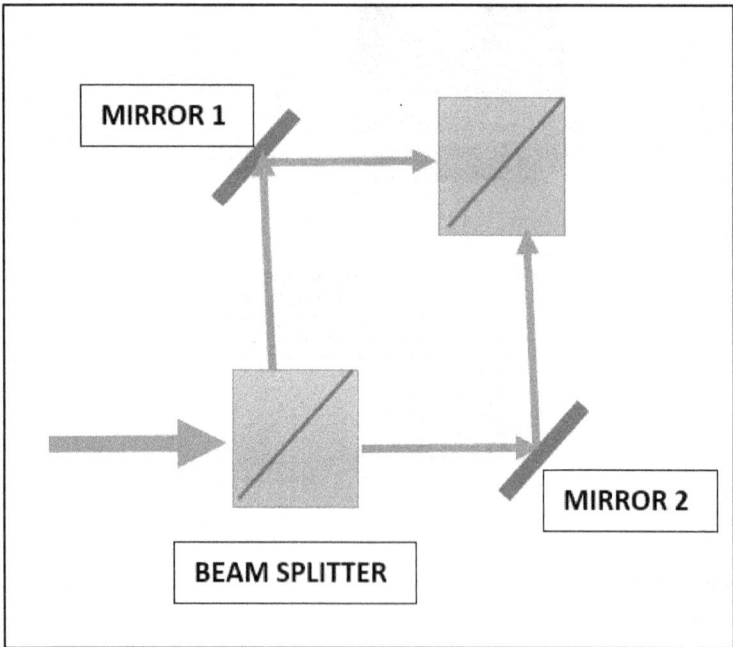

Figure 13.7 Principle of holography.

A coherent source of light like laser helps in data storage by creating three-dimensional images by the interference of light beams. This is the basic principle of a hologram which is a potential technology of high data storage. This is in the same way as the traditional optical discs in which the data was stored on the surface of the disk. The working starts when the coherent source of the light beam (laser beam) is split into two beams—the data-carrying beam and the signal beam. This signal beam travels straight, bounces off one mirror and propagates through a shutter system which regulates passing and blocking lights at respective zeros and ones. The reference beam (second beam) emerges out of the beam splitter and follows a new path towards the crystal. An interface pattern is created at the point at which the two beams and this results in the formation of a 3-dimensional image called **hologram** which is recorded in a light-sensitive storage medium.

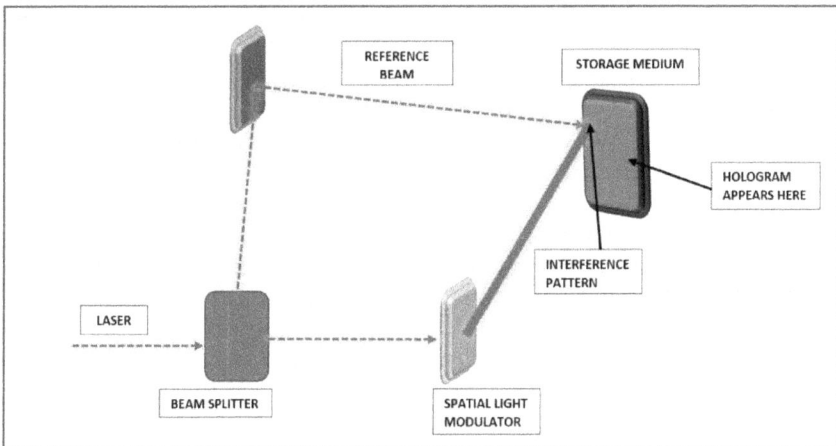

Figure 13.8 Formation of a hologram.

If we change the beam angle, media, and wavelength, a variety of holograms can be stored in a single volume. In order to read the stored data, the same reference beam is used, and it is casted into the crystal at exactly the same beam angle at which it was entered to store the data. The reference beam is diffracted from the crystal in order to recreate the original page that has been stored and projected onto a detector that interprets and transmits the digital data to the computer. This type of Holographic Storage is based on WORM (Write Once Read Many) technology. According to an estimate over 4.4 million individual pages can be stored on a holographic drive.

References

1. Myers, W. (1976) Key developments in computer technology: A survey. *Computer*, **9**(11), 48-77.
2. Psaltis, D., and Burr, G. W. (1998) Holographic data storage. *Computer*, **31**, 52–60.
3. Broer, D. J., and Vriens, L. (1983) Laser-induced optical recording in thin films. *Applied Physics A*, **32**, 107-118.
4. Gu, M., and Li, X. (2010) The road to multi-dimensional bit-by-bit optical data storage. *Optics and Photonics News*, **21**, 28–33.
5. Li, X., Chon, J. W., Wu, S., Evans, R. A., and Gu, M. (2007) Rewritable polarization-encoded multilayer data storage in 2,5-dimethyl-4-(p-nitrophenylazo)anisole doped polymer. *Optics Letters*, **32**, 277–279.
6. Alexoudi, T., Kanellos, G. T., and Pleros, N. (2020) Optical RAM and integrated optical memories: a survey. *Light: Science and Applications*, **9**, 91-104.
7. Tufte, O. N., and Chen, D. (1973) Optical techniques for data storage. *IEEE Spectrum*, **10**, 26-32.
8. Strehlow, W. W., Dennison, R. L., and Packard, J. R. (1974) Holographic data Store. *Journal of the Optical Society of America A*, **64**, 543.
9. Matick, R. E. (1972) Review of current proposed technologies for mass storage systems. *Proceedings of the IEEE*, **60**, 266-289.
10. Chen, D. (1974) Magnetic materials for optical recording. *Applied Optics*, **13**, 767-778.
11. Rajchman, J. A. (1970) An optical read-write mass memory. *Applied Optics*, **9**, 2269-2271.

Chapter 14

Switches and Sensors

Introduction

A **solid-state Switch** is also known as a **solid-state relay (SSR)** and it comprises of a solid-state material which has features are better than the traditional switches/relays. They are employed for switching of power automatically. The relays are more commonly used nowadays to get rid of the typical manual switching techniques that are difficult to handle. It is a non-contact switch and is made of a solid-state element which can control high load current with a smaller control signal. A solid-state relay is highly safe since there is no possibility of sparking due to the characteristics of the solid-state element used in the relay. The switching OFF and ON state in the solid-state relay is easy to achieve. Solid-state relays have different properties which make them unique:

1. Solid-state relays lack mechanical elements.
2. There are very low voltage rises.
3. Solid-state relays use zero-crossing technology

Solid-state relays have other advantages like corrosion resistance, Vibration resistance, high reliability and long life. The main feature of the solid-state relays is the higher load capacity.

14.1 Types of Switching Devices

Transistor: The transistor is also termed as **MOSFET** (metal-oxide-semiconductor field-effect transistor) indicating that it is a semiconductor device made up of two metal-oxide semiconductor field effect transistors (MOSFETs), one N-type and one P-type, integrated on a single silicon chip. It is commonly used for switching Direct Current (DC) loads.

SCR: SCR stands for the **silicon-controlled rectifier**. It controls current flow and is a four-layer solid-state device. It acts as a switch, conducting when its gate receives current, and it continues to conduct for as long as it is forward biased. It is preferably used for switching all types of Alternating Current (AC) loads.

TRIAC: A TRIAC stands for **triode for alternating current** and is an electronic component equivalent to two silicon-controlled rectifiers (SCR) joined in parallel but with the reversed polarity (inverse parallel) and with their gates connected together. This makes it a bidirectional electronic switch that can conduct current in either direction. It is ideal for switching resistive Alternating Current (AC) loads.

14.2 Variables of Solid-State Switch/Relays

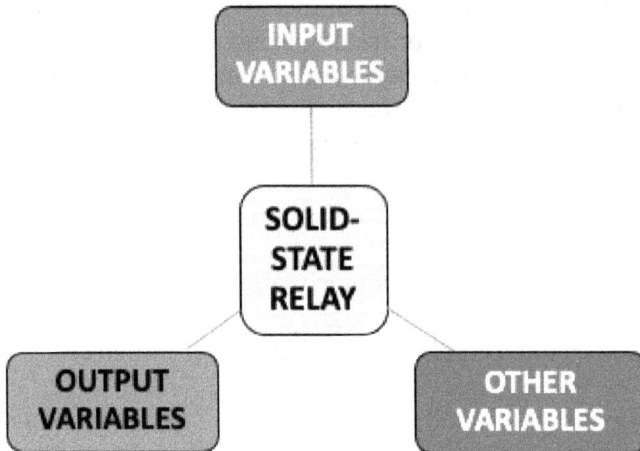

Figure 14.1 Classification variables of solid-state relay.

There are certain variables for solid-state switches, and they are categorized as input variables, output variables and other variables.

Input Variables

Input Current and Input Voltage Range: The input current means the input current value at a particular input voltage and input voltage range for a solid-state relay is the range of voltage between minimum and maximum required for the relay to function normally at room temperature.

Switch-On Voltage and Switch-Off Voltage: The switch-on voltage, when the voltage at the input terminal is greater than or equal to the switch-on voltage, the output terminal will be turned on. Switch-off voltage, when the voltage at the input terminal is less than or equal to the shutdown voltage, the output terminal will be turned off.

Zero-Crossing Voltage: It is a voltage range that is estimated by the components of the zero-crossing switch. Its value is either very low or negligible. If the zero-crossing voltage is higher than the power supply voltage, then the zero-crossing switch is OFF, however, in the reverse condition, the zero-crossing switch is ON.

Output Variables

Rated Output Voltage: The maximum load operating voltage that the output terminal can stand up to is called the rated output voltage.

Output Voltage Drop: The resolute output voltage at the rated operating current, when the solid-state switch is in ON state, is called the output voltage drop. It is a criterion of the quality and performance of solid-state switches/relays. That solid-state switch is termed to be better which has a smaller output voltage drop.

Overload Current: Overload current is also termed as inrush current or input surge current. When the solid-state switch is in ON-condition, the maximum current value that does not damage the device and can be tolerated by the output terminals is the overload current.

Other Variables

Power Consumption: The maximum amount of power that a solid-state switch consumes in ON and OFF states is known as power consumption.

Switching-on Time / Switching-off Time: This is the time delay observed during turning-on and turning-off of the solid-state switch is termed as the switching-on time and switching-off time. The performance of a solid-state switch/relay depends on this time. The shorter the duration of this time, the better is the performance of the solid-state switch/relay.

Insulation Resistance: The stated value of resistance between the input and output terminal of a solid-state switch is referred to as the insulation resistance.

Operation Temperature: Operation temperature is the normal working environment temperature range allowed for the solid-state relay/switch.

14.3 Principle of Solid-State Switch/Relay

There are two types of solid-state switch/relay: alternating current (AC) solid-state switch/relay and direct current (DC) solid-state switch/relay. The AC solid state switch/relay has input and output terminals. On providing a specific control signal to the input terminals of the solid-state switch/relay, the ON and OFF function on the output terminal is performed respectively. With the help of the coupling circuit in the solid-state switch/relay, a passage is formed between the input and output of the switch/relay.

The cut-off function is also brought about by this coupling circuit, in case of the unpredicted circumstances which are set initially before installing the relay. Optical couplers are used in the coupling circuit of the solid-state switch/relay as they have commendable sensitization, high speed response input-output insulation level is also good. LED (light emission diode) is used to match the level of the input signal in the load used in the solid-state switch/relay. The output is either "0" or "1" in the solid-state relay and it is connected to a computer for interface.

In the case of the DC solid state switch/relays, there is no snubber circuit and zero-crossing control circuit. A transistor with a larger value is used for the switching of the solid-state relay. Other working principle for this relay is the same as the AC working principle.

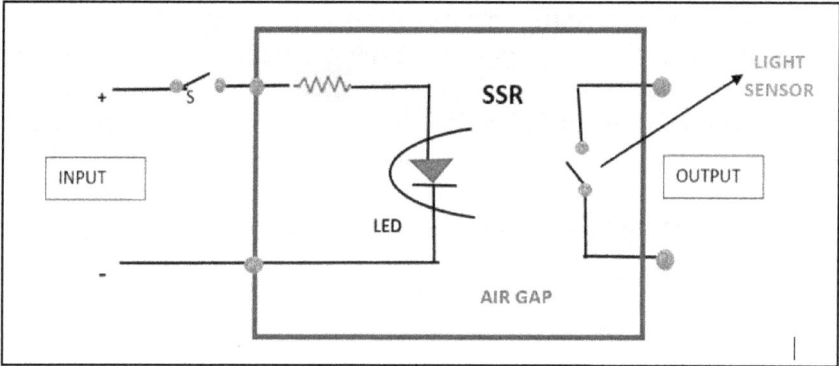

Figure 14.2 General structure of a solid-state switch/relay.

14.4 Working of Solid-State Switch/Relay

Solid-state switch/relay has two input terminals (A & B) and two output terminals (C & D), it is a four-terminal active device.

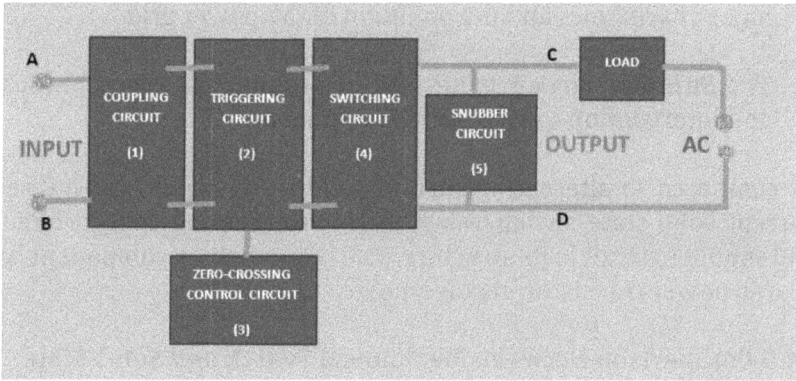

Figure 14.3 Working design of solid-state switch.

In the working of a solid-state switch, there is a control signal between A and B (input terminal) that manages ON-OFF state between C and D (output terminal) thereby completing the switching function. The functioning of various components is as under:

a) The **coupling circuit** is a pass over between input and output terminal. Its role is to supervise the signal that is being put into from A and B but terminate the electrical circuit between input and output thus safeguarding the input from output. The structure of coupling circuit consists of light emitting diode which have unique properties of excellent sensitivity, response speed and insulation at input and output. This is directly linked to a computer interface.

b) The **triggering circuit** engenders a triggering signal so that the switching circuit starts working.

c) The **zero-crossing control circuit** prevents the switching circuit from producing radio frequency interface (RFI) which pollutes the power grid as high harmonics. Zero crossing indicates that when solid-state switch/relay is in on-state, there is placed a control signal facilitating the alternating current voltage in crossing zero; on switching off the control signal, Solid-State Switch/Relay does not come in off-state until AC current is at the junction of zero potential. Such an arrangement prevents the interference of higher harmonics and the pollution of the power grid.

d) The **Snubber circuit** is devised to prohibit the reverberations and intervention to switching component.

As compared to alternating current solid-state switch/relay, direct current solid-state switch/relay lacks zero-crossing control circuit and snubber circuit in its structure, also the switching component has a large power transistor; rest is similar.

14.5 Comparison Between Mechanical Switch and Solid-State Switch

There are certain important differences between mechanical and solid-state relays which are:

1. Normal relays have mechanical elements in them whereas solid-state relays have no mechanical elements.
2. The load capacity for the solid-state relay is higher than that of the normal relay.

3. Moving part crossing is used in normal relays whereas zero crossing is used in the solid-state relays for its operation.
4. The elements used in solid-state relays are long-lasting and efficient as compared to normal relay.

14.6 Applications of Solid-State Switch/Relays

The most common application of the solid-state relay is the load switching of alternating current load. A number of household circuits have solid state relays for AC load switching. Solid state relays have different applications with the power control, moreover, different solid-state relays are used for heater controls.

14.7 Sensors

A sensor is a device which senses, detects or measures a physical property and records, indicates or responds to it. In the structure, between a sensor and an actuator, a signal processing unit controls the whole system. Because the signal processing unit can be built as a semiconductor device, it is desirable to have the sensor device and the signal processing unit on the same chip. The raw signal available from the sensors are then passed through a signal processing unit which eliminates undesired signals thus making the desired signal strong through a preamplifier stage. The role of actuator is to accept an electrical input and convert it into physical action.

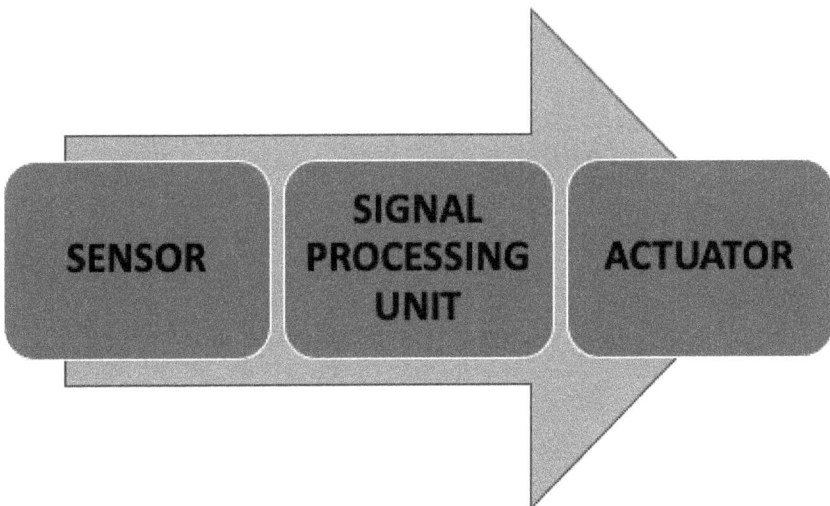

14.8 Types of Sensors and Their Functionality

Light Sensors

Infra-Red Sensor- It is also called as Infra-Red Transmitter and is used as a source of Infrared rays. The range of these transmitter frequencies are greater than the microwave frequencies. A photodiode is connected to infrared LED (light emitting diode) to sense the rays generated by it. Therefore, the pair of IR LED and photodiode is called **IR sensor.**

Photodiode- Photodiode is a semiconductor device which is used as an IR Receiver used to detect the light rays. The construction of photodiode is similar to the normal p-n junction diode however its working is somewhat different. The property of a p-n junction to allow small leakage currents when it is in reverse bias is helpful in detecting the light rays. A photodiode is structured in a way that when high intensity light rays fall on the p-n junction, the leakage current increases. Thus, a photodiode can be used to sense the light rays and maintain the current through the circuit.

Light Dependent Resistor (LDR)- Light Dependent Resistor is a resistor that depends upon the light intensity. It works on the principle of photoconductivity i.e., conduction due to the light. It is generally made up of cadmium sulfide. When light falls on the LDR, the resistance decreases. If light is not focused on the surface of LDR, then it will not respond.

Temperature Sensors

Thermistor- A thermistor has a negative value for coefficient of temperature meaning that temperature increases on decreasing the resistance and thus it is used to detect the variation in temperature. So, the thermistor's resistance can be varied with the rise in temperature which causes more current flow through it. This change in current flow can be used to determine the amount of change in temperature. Thermistor is used to detect the rise in temperature and control the leakage current in a transistor circuit which helps in maintaining its stability.

Thermocouple- Thermocouple is also used to detect the variation in temperature. Here two metals join together and form an intersection. The potential across the terminal of a thermocouple varies when the intersection of two different metals is exhibited to high temperature. This varying potential is used to read the amount of change in temperature.

Pressure/Force/Weight Sensors

Strain Gauge- The change in pressure on application of a load is detected by a Strain Gauge. These are particularly used to find out the amount of pressure that the wings of an airplane can stand and also to calculate the number of vehicles that can be allowed on a particular road. There is a wire arranged in a zig-zag pattern on a limber board and when pressure is registered on that board, the wire bends in the direction that causes a change in length and area of cross-section of the wire. This change in turn changes the resistance if the wire which can be measured by a Wheatstone Bridge. In one of the four arms of the bridge, the strain gauge is placed while other values remain the same. Thus, on applying pressure, resistance changes varying the current passing through the bridge and hence pressure can be calculated.

Position Sensors

Potentiometer- A potentiometer detects the position. There are various resistors linked to different poles of the switch. A potentiometer can be either rotary or linear type. The rotary type potentiometer, has a wiper that is connected to a long shaft which can be rotated. On rotating the shaft, the position of the wiper changes in a way that the total resistance alters causing the change in the output voltage which can be calibrated to assess the change in its position.

Sound Sensors

Microphone- Microphone can detect the audio signal and convert them into small electrical current. There is a diaphragm which vibrates when a sound wave hits it and this in succession moves a magnet near a coil. Also, there are condenser microphones, working on the principle of capacitance. There are parallel conducting plates in a capacitor which store charge. When there is sound coming into a condenser microphone, one plate of the capacitor vibrates, and the altering capacitance is converted into electrical current.

Touch Sensors

Resistive Touch Sensor- In a resistive touch screen, there is a sandwich of base resistive sheet, air gap and a conductive sheet beneath the screen. A small is applied to the sheets. On touching the screen, contact is established between the conductive sheet and the resistive sheet resulting in current flow at that point. The computer program senses this location and executes the relevant action.

Capacitive Touch Sensor- In capacitive touch screen, there is an electric field throughout the screen. On touching the screen, a circuit is formed due to electrostatic charge that flow through our body. Then, the computer program senses this location and executes the relevant action.

Gas Sensors

They are helpful in detecting the gas leakage in industry. Below a metal sheet, there is a sensing element which is connected to the terminals where a current is applied to it. When the gas particles hit the sensing element, it leads to a chemical reaction such that the resistance of the elements varies and current through it also changes which finally can detect the gas.

References

1. Wenyu, Z., Erica, D., William R. D., Lei, F., Ali, T,. John-Carl, O., Diego, B., James, R. H., and and Stoddart, J. F. (2011) A solid-state switch containing an electrochemically switchable bistable poly[n]rotaxane. *Journal of Materials Chemisty*, **21**(5), 1487-1495.
2. Jonathan, P. H., and Ciszek, J. W. (2017) Solid state and surface effects in thin-film molecular switches. *Photochemical and Photobiological Sciences*, **16**, 1095-1102.
3. Wang, L., Zhang, D., and Wang, Y. (2016) High performance solid-state switches using series-connected SiC-MOSFETs for high voltage applications, *IEEE 8th International Power Electronics and Motion Control Conference (IPEMC-ECCE Asia)*, pp. 1674-1679.

4. Verma, N., Gupta, K., and Mahapatra, S. (2015) Implementation Of solid state relays for power system protection. *International Journal of Scientific & Technology Research*, **4**(6), 65-70.

5. Vinod, M., Devadasan, S. R., Rajanayagam, D., Sunil, D. T., and Thilak, V. M. M. (2018) Theoretical and industrial studies on the electromechanical relay. *International Journal of Services and Operations Management*, **29**(3), 312 – 331.

6. Lihuiyan, P. M. (2016) Precise control of process temperature using AI-7048 temperature controller and solid-state relay (SSR) control system. *International Journal of Science and Research*, **5**(9), 1238–1242.

7. De Rooij, N. F. (1989) Chairman Transducers '89, *5th International Conference on Solid-State Sensors and Actuators*, Switzerland.

8. Gauthier, M., and Chamberrand, A. (1977) Solid-state detectors for the potentiometric determination of gaseous oxides. *Journal of Electrochemical Society*, **124**, 1579–1583.

9. Heiland, G. (1982) Homogeneous semiconducting gas sensors. *Sensors and Actuators*, **2**, 343–361.

10. Ajmera, P. (2017) A review paper on infrared sensor. *International Journal of Engineering Research & Technology*, **5**(23), 284-289.

11. Murtala, A., Haruna, I., Paul, O., and Ahmad, A. S. (2013). An electronic switch sensor with a point-to-point intrusive monitoring system. *International Journal of Information Technology, Modelling and Computing*, **1**, 87-98.

12. Shaikh, A., and Pathan, S. (2012) Research on wireless sensor network technology. *International Journal of Information and Education Technology*, **2**(5), 476-479.

13. Kumar, R., and Kumar, R. (2016) Review paper on wireless sensor networks. *International Journal of Engineering Research & Technology*, **4**(32), 92-97.

14. Prasad, L., Iyengar, S. S., Kashyap, R. L., and Madan, R. N. (1991) Functional characterization of sensor integration in distributed sensor networks. *IEEE Transactions on Systems, Man, and Cybernetics*, **21**(5), 1082-1087.

Index

A

absolute zero, 75, 78
additives, 15, 92
agate, 18
amorphous, 1-2, 85, 97
anionic vacancies, 11
anisotropic, 2
antiferromagnetic
materials, 56-57
arc discharge method, 89
artificial photosynthesis,
108-109, 111, 113
atomic absorption
spectroscopy, 24
attrition, 39

B

ball mill, 19-20
batteries, 50, 59
BCS theory, 76, 78
biosensors, 50
bipolaron, 45-46
blu-Ray, 119
bridge rectifiers, 98

C

C60 fullerene, 86-87
C70 fullerene, 87
cache, 116

N

O

P

www.ingramcontent.com/pod-product-compliance
Lightning Source LLC
Chambersburg PA
CBHW071649210326
41597CB00017B/2159